Political and Investment Risk in the International Oil and Gas Industry

Political and Investment Risk in the International Oil and Gas Industry

Steven A. Mucci

LEXINGTON BOOKS
Lanham • Boulder • New York • London

Published by Lexington Books
An imprint of The Rowman & Littlefield Publishing Group, Inc.
4501 Forbes Boulevard, Suite 200, Lanham, Maryland 20706
www.rowman.com

Unit A, Whitacre Mews, 26-34 Stannary Street, London SE11 4AB

British Library Cataloguing-in-Publication Information Available

Library of Congress Cataloging-in-Publication Data Available

ISBN 978-1-4985-4612-6 (cloth: alk.paper)
ISBN 978-1-4985-4613-3 (electronic)

∞™ The paper used in this publication meets the minimum requirements of American National Standard for Information Sciences—Permanence of Paper for Printed Library Materials, ANSI/NISO Z39.48-1992.

Printed in the United States of America

Contents

List of Figures

List of Tables

Preface

The idea for a one-volume treatment of the evolution of political and investment risk management in the oil and gas industry developed during my pursuit of a PhD at the University of Missouri in St. Louis. When I began this project, in August 2014, the price of West Texas Intermediate crude oil (WTI) was $106.91 per barrel; crude oil demand was high; and supply appeared limited by technical, economic and political factors. However, by February 11, 2016, the price of WTI had reached a thirteen-year low of $26.19 per barrel; supply temporarily exceeded demand; and world crude oil inventories had risen substantially. At the time of writing, (April 28, 2017) the price of WTI closed at $49.33 per barrel.

The absolute price of oil does not have a significant effect on political and investment risk, but the rate at which the price of oil increases relative to its long-term price trend does have a significant effect on political and investment risk, specifically the risk of expropriation. When crude oil prices are rising rapidly relative to the long-term price trend, the incentive for and political pressure on national governments to expropriate oil and gas assets increases. In addition, from 1970 to the present, the prices of oil and gas, and the rate at which they have increased or decreased have been erratic, making the risk of expropriation an ongoing concern.

For example, between 1971 and 1979, the average global price of crude oil increased from $2.24 per barrel to $31.61 per barrel (1,311%). Between 2004 and 2011, the average global price of crude oil increased from $38.27 per barrel to $111.26 per barrel (191%). In each period, the number of expropriations increased substantially. From 2004 to 2011, major expropriations took place in Venezuela, Russia, Argentina, Ecuador and Bolivia. The rate at which the price of oil increases or decreases is influenced by political and economic events that are difficult to predict. It is possible that another rapid increase in the price of oil will occur and that oil and gas assets owned by foreign investors will be expropriated. For these reasons, the management of political and investment risk continues to be important. This book examines the evolution of political and investment risk management and evaluates how effective it has been.

NOTE

1. Sergei Guriev, Anton Kolotilin and Konstantin Sonin, "Determinants of Nationalization in the Oil Sector: A Theory and Evidence from Panel Data," *Journal of Law, Economics & Organization* 27 (2) (2011) p. 301.

Acknowledgments

This book would not exist without the encouragement I received from professors, Kenneth Thomas, J. Martin Rochester, David B. Robertson and David Kimball at the University of Missouri, St. Louis. Each of them was instrumental in giving me the confidence to proceed with this project. My editors at Lexington Books, Madhumitha Koduvalli and Joseph Parry, were indispensable in transforming my original manuscript into the book it has become.

List of Abbreviations

AAR	Alfa-Access-Renova
ACES	Alaska's Clear and Equitable Share
ADB	Asian Development Bank
ADNOC	Abu Dhabi National Oil Company
AIDB	African Development Bank
AIG	American International Group
AIOC	Anglo-Iranian Oil Company
AMMA	Australian Mines and Metals Association
ANP	Brazilian National Petroleum Agency
AP	Associated Press
APEC	Asia-Pacific Economic Cooperation
ARCO	Atlantic Richfield Company
AzBTC	Azerbaijan BTC Limited
BADEA	Arab Bank for Economic Development in Africa
BBC	British Broadcasting Corporation
BIT	bilateral investment treaty
BLEU	Belgium-Luxembourg Economic Union
BTC	Baku-Tbilisi-Ceyhan
CAPEX	capital expenditures
CBC	Canadian Broadcasting Corporation
CDB	China Development Bank
CEO	chief executive officer
CFF	Carbon Finance Facility
CFIUS	Committee on Foreign Investment in the United States
CLC	International Convention on Civil Liability for Oil Pollution Damage
CNN	Cable News Network

CNOOC	China National Offshore Oil Corporation
CNPC	China National Petroleum Corporation
CPC	Communist Party of China
CRCICA	The Cairo Regional Centre for International Commercial Arbitration
DPR	Department of Petroleum Resources
E&Y	Ernst and Young
ECA	export credit agency
ECT	Energy Charter Treaty
EIA	Energy Information Administration
EPA	Environmental Protection Agency
EPSA	Exploration and Production Sharing Agreements
ESMAP	Energy Sector Management Assistance Program
EU	European Union
Eurostat	Statistical Office of the European Communities
FASB	Financial Accounting Standards Board
FDI	foreign direct investment
FET	Fair and Equitable treatment clause
FNC	Friendship, Commerce, and Navigation treaties
FPSO	floating production storage and offloading facility
FSB	Federal Security Service of the Russian Federation
GDP	gross domestic product
GEF	Global Environment Facility
H&P	Helmerich & Payne
IASB	International Accounting Standards Board
IBRD	International Bank for Reconstruction and Development
ICBC	Industrial and Commercial Bank of China
ICC	International Chamber of Commerce
ICSID	International Centre for the Settlement of Investment Disputes
ICSID/AF	ICSID Additional Facilities
IDA	International Development Association
IDB	Inter-American Development Bank
IEA	International Energy Agency
IEF	International Energy Forum
IFC	International Finance Corporation
IIA	international investment agreement
IMF	International Monetary Fund
IOC	international oil company
IPO	initial public offering
ISIL	Islamic State in Iraq and the Levant
JBIC	Japan Bank for International Cooperation
JODI	Joint Oil Data Initiative

JV	joint venture
KRG	Kurdistan Regional Government
LCIA	London Court of Arbitration
LNG	liquefied natural gas
LPG	liquefied petroleum gas
MAI	Multilateral Agreement on Investment
MBPD	million barrels per day
MDOC	Midland Oil Company
MFN	most favored nation
MIGA	Multilateral Investment Guarantee Agency
MLP	Master Limited Partnership
MOC	Missan Oil Company
NAFTA	North American Free Trade Agreement
NASDAQ	National Association of Securities Dealers Automated Quotations
NEA	National Energy Administration
NGL	natural gas liquids
NIE	new institutional economics
NIOC	National Iranian Oil Company
NNPC	Nigerian National Petroleum Corporation
NOC	national oil company
NOC	North Oil Company
NPR	National Public Radio
NRDA	Natural Resource Damage Assessment
NYSE	New York Stock Exchange
O&GJ	Oil & Gas Journal
OECD	Organization for Economic Cooperation and Development
OIL	Oil Insurance Limited
OLADE	Latin American Energy Organization
ONGC	Oil and Natural Gas Company Limited
OPA	Oil Pollution and Control Act
OPEC	Organization of Petroleum Exporting Countries
OPIC	Overseas Private Investment Corporation
OSCE	Organization for Security and Cooperation in Europe
PAN	National Action Party
PCA	Permanent Court of Arbitration
PDVSA	Petroleos de Venezuela
PEMEX	Petroleos Mexicanos
PIB	Petroleum Industry Bill
PRI	political risk insurance
PSA	production-sharing agreement
PSC	production-sharing contract

PTP	publicly traded partnership
PWYP	Publish What You Pay U.S.
SAIC	Science Applications International Corporation
SCC	Stockholm Chamber of Commerce Arbitration
SEC	Securities and Exchange Commission
SIC	Standard Industrial Classification
SINOPEC	China Petroleum and Chemical Corporation
SOC	South Oil Company
SOCAR	State Oil Company of Azerbaijan Republic
SPE	special purpose entity
SPE	Society of Petroleum Engineers
TAPS	Trans-Alaska Pipeline System
TPAO	Türkiye Petrolleri Anonim Ortaklığı
U.S.	United States
U.S. GAAP	U.S. Generally Accepted Accounting Principles
UAE	United Arab Emirates
UK	United Kingdom
UNCITRAL	United Nations Commission on International Trade Law
UNCTAD	United Nations Conference on Trade and Development
UNSD	United Nations Statistical Division
WAMUA	Western Australia Maritime Union of Australia
WO	World Oil
WPC	World Petroleum Council
WTI	West Texas Intermediate
WTO	World Trade Organization
YFPB	Yacimientos Petroliferas Fiscales Bolivianos
YPF	Yacimientos Petrolíferos Fiscales

Introduction

THE PROBLEM

Measuring and predicting political risk is difficult. For example, Venezuela actively encouraged foreign direct investment in the oil and gas industry until 1999, but after Hugo Chavez was elected president of Venezuela, the investment climate became increasingly hostile. Vladimir Putin's drive for renationalization of the Russian oil and gas industry did not begin until 2005, even though Putin was elected prime minister in 1999. The government of Argentina did not expropriate and nationalize Repsol's 57.4 percent interest in Yacimientos Petrolíferos Fiscales (YPF) until 2012. Even though Nestor Kirchner had served as president of Argentina from May 25, 2003 to December 10, 2007, and was considered by some to be a left-wing president, he did not propose left-wing policies such as socialization of production or the nationalization of public services. On December 10, 2007, his wife, Cristina Fernandez de Kirchner, succeeded him and it was during her presidency that Repsol's interest in YPF was expropriated.

The popular uprisings of the Arab Spring which began on December 10, 2010, caught almost all political analysts by surprise, as did the chaos that followed in Libya, Tunisia, Yemen, Bahrain and Syria; and the sectarian and ethnic divisions that were apparent before the United States occupied Iraq and were exacerbated after the U.S. military withdrew from Iraq, have proven to be more disruptive and intractable than the coalition partners anticipated.

Financial and economic crises continue to be a recurring feature of the global economy. The global financial crises that began in 2007 and continued through 2009, required extraordinary measures by central banks in the developed countries to avert an even worse global recession. Consequently, after a period of relatively low investment claim losses, there is a renewed awareness that investment risk, driven by political events, is still difficult to predict.[1]

This is particularly true in the international oil and natural gas industry, in which there is often a shift in bargaining power over the life of a project between the nation supplying the natural resource, the transit countries through which the transport pipeline passes and the multinational companies providing the financial capital and technical expertise needed to find and produce the oil and gas. Until exploration is completed and the field development facilities and pipeline infrastructure are in place, the superior bargaining power lies with the multinational companies, because the host government often does not have the financial, technical and marketing resources needed to find and produce the oil and natural gas. However, after the investment in the facilities and export infrastructure have been made by the foreign investors, the bargaining power shifts to the host government and the transit countries, because they have the power to interrupt operations, pass legislation and impose regulations that have an adverse effect on the foreign investor.[2]

Furthermore, oil and gas projects often extend over several decades. During that time, the host government's economic policy, political ideology and need for revenue may change. Consequently, what the host government and transit countries previously considered attractive financial terms, may no longer satisfy their needs, resulting in impairment of the foreign investor's assets, creeping expropriation or outright nationalization. Raymond Vernon described this phenomenon as an "obsolescing bargain."[3] The new regime may demand larger financial returns than originally agreed on, in what have become, mature, low-risk projects. The demands usually take the form of forced contract renegotiation, increased fiscal take or expropriation.[4]

The inclination of host governments to demand more favorable terms are observed not only in developing countries. For example: (1) the state of Alaska has increased the financial burden levied on oil investors several times since the commissioning of the Trans-Alaska Pipeline System (TAPS) in 1977. More recently, the Alaska's Clear and Equitable Share (ACES), a fiscal system implemented in 2007, significantly increased the fiscal take from the oil companies.[5] (2) The government of the United Kingdom increased the corporate tax on North Sea oil producers (termed supplemental corporation tax) three times between 2002 and 2011, placing an increased fiscal burden on oil companies operating mature oilfields with declining production. In the Organization for Economic Cooperation and Development (OECD) countries, asset expropriation is rare and tax increases are usually not made on a retroactive basis, but in the developing world, forced acceptance of onerous financial terms by investors, sometimes retroactively, and outright expropriation were still common between 2000 and 2010, for example, in Algeria, Ecuador, Kazakhstan, Russia, Argentina and Venezuela.[6] The consequences of political and investment risk can be mitigated to varying degrees by the development of appropriate institutions.

IMPORTANCE OF THE PROBLEM

The effectiveness of oil and gas companies in managing political and investment risk is important for several reasons. First, it is important to the profitability of individual firms, and the perceived and real level of risk affects the level of exploratory and development expenditures. Viewed from a global perspective, oil and natural gas are essential for sustaining current levels of economic activity and making economic growth possible. Second, the balance between supply and demand determines the price of oil and natural gas, and the price of oil and natural gas affects the economies of all nations. Therefore, the price and availability of oil and natural gas are also matters of national security. Third, the search for a secure supply of oil and gas affects the political, military and economic relations between countries. Consequently, every developed and developing country has placed energy policy at or near the top of its national priorities.

Supply and Demand

World consumption of petroleum liquids (crude oil and liquefied petroleum gas [LPG]), has increased from 83.107 MBPD (million barrels per day) in 2004 to 92.086 MBPD in 2014.[7] The consumption of petroleum liquids is expected to increase from 92.086 MBPD in 2014 to 103.5 MBPD in 2040. The production of petroleum liquids increased from 80.938 MBPD in 2004 to 88.673 MBPD in 2014.[8] The production of petroleum liquids is expected to increase from 88.673 MBPD in 2014 to 103.5 MBPD in 2040.[9] This assumes that the supply and demand for petroleum liquids comes back into balance at $80.00 per barrel in 2020.[10] The Organization of Petroleum Exporting Countries (OPEC) forecasts that the world price of crude oil will reach $95.00 per barrel in 2040.[11]

Capital Expenditures

Global capital expenditures and exploration expense (CAPEX) in the exploration and production sector of the oil and gas industry were $688 billion in 2014, but decreased by 27 percent to $502 billion in 2015.[12] This decrease in CAPEX was the result of the decrease in the world price of crude oil, for example, the price of West Texas Intermediate (WTI) decreased from $106 per barrel in June 2014, $56 per barrel at the end of December 2014 and $29.71 per barrel on February 8, 2016. Global CAPEX was expected to decline further in 2016 to $426 billion.[13] CAPEX is forecast to increase by 5 percent to $448 billion in 2017.[14]

The International Energy Agency (IEA) cautioned against complacency because of the oversupply of crude oil in early 2016. The IEA estimates that to compensate for declining production in the existing fields and to keep future output at current production levels global CAPEX will need to average $630 billion per year.[15]

The large amount of financial and physical capital required in oil and natural gas exploration and the immobility of these assets make them particularly vulnerable to the obsolescing bargain. Furthermore, when the risk of contract renegotiation or expropriation appears to be increasing, investments in oil and natural gas that would otherwise be made will not be made, thereby reducing the supply and increasing the price of oil and natural gas. However, the development of appropriate institutions can reduce or reallocate this risk. Understanding how these institutions can and do reduce risk is important to countries, international oil companies (IOCs) and national oil companies (NOCs) for several reasons.

Specific Issues in the International Oil and Gas Industry

First, the recurrence of resource nationalism in democratic oil exporting countries like Mexico and Venezuela, and the nationalist policies of more authoritarian regimes like Russia to retain or regain control of their oil and gas resources is a continuing threat to direct foreign investment and reduces the incentive for IOCs to explore for oil and natural gas.

Second, IOCs possess technology that could be used successfully in countries that rely almost exclusively on their NOCs for exploration and production technology, but oil-producing countries and their NOCs have been reluctant to allow IOCs to explore in their countries. Consequently, the IOCs have been forced to explore in more difficult environments, for example, ultra-deep water and the Arctic Circle where exploration and production costs and technological risk are much higher. However, the commercial feasibility of fracking has given the U.S. domestic oil producers an alternative to conventional drilling.

Third, many national oil companies, for example, Petroleos Mexicanos (PEMEX) in Mexico, Petroleos de Venezuela (PDVSA) in Venezuela and the National Iranian Oil Company (NIOC) in Iran have not reinvested enough of their oil and gas revenue in the discovery of new oil and natural gas reserves to replace the reserves that are being depleted.

Fourth, many NOCs are responsible for the funding and operation of social programs in their country and the exploration and production of oil and gas. The inability of most NOCs to focus exclusively on commercial objectives and the absence of a true bottom line have frequently led to the mismanagement of financial resources.

Fifth, the lack of transparency in the operation and financial performance of many national oil companies makes private sector lenders reluctant to lend to these NOCs, consequently not all sources of capital are available to national oil companies.

Sixth, international law provides limited enforcement mechanisms as a deterrent to the unilateral abrogation of agreements between host countries and international oil companies and to the expropriation of assets owned by IOCs in developing countries.

PURPOSE OF THE BOOK

The purpose of this book is to provide oil and gas industry managers, academics and students a one-volume explanation of how political risk management tools have evolved, how they are used and how effective they have been in the oil and gas industry. For many oil managers and executives who have responsibilities other than risk management, this book will provide a better understanding of what their colleagues in the risk management, legal and finance departments do to protect the company's investments. Second, it will provide academics with useful background for their own research in public policy, energy regulation and institutional economics. Third, it will provide students interested in energy issues with a starting point for understanding the industry and how it operates.

PLAN OF THE BOOK

Chapter 1 begins by defining risk, examining the differences between portfolio investment and foreign direct investment and outlining the key questions answered in subsequent chapters. Chapter 2 provides a description of the three principal types of organizations operating in the international oil and gas industry (national governments, national oil companies and international oil companies). Chapter 3 examines the most common contract provisions in oil and gas contracts; the relationship between domestic courts, foreign courts and arbitral tribunals; common provisions in bilateral investment treaties; legal systems and disputes; the relationship between oil consumption, foreign direct investment and bilateral investment treaties; and the causes and frequency of expropriation. Chapter 4 examines the principal sources of financing in the oil and gas industry, available financing structures, fiscal regimes, alliances, managing commercial and noncommercial risk, financial reporting and operational transparency, financial reporting quality and the estimation and valuation of oil and gas reserves. Chapter 5 examines multilateral treaties, international energy forums, financial reporting rule making institutions, signature bonuses, energy diplomacy and foreign direct investment and national security. Chapter 6 contains six case studies. The book concludes with a summary and conclusions.

NOTES

1. Multilateral Investment Guarantee Agency, *World Investment and Political Risk*, World Bank Group (Washington, DC 20433, 2012), p. 42.

2. David Wood, *Petroleum Economics, Risk and Opportunity Analysis*, Chapter 10 in Betty J. Simkins and Russell E. Simkins, Energy Finance and Economics – Analysis and Valuation, Risk Management, and the Future of Energy (Hoboken, NJ: John Wiley & Sons, Inc., 2013), p. 243.

3. Raymond Vernon, *Sovereignty at Bay: The Multinational Spread of US Enterprises*, (New York: Basic Books, 1971).

4. David Wood, *Petroleum Economics, Risk and Opportunity Analysis*, Chapter 10 in Betty J. Simkins and Russell E. Simkins, Energy Finance and Economics – Analysis and Valuation, Risk Management, and the Future of Energy (Hoboken, NJ: John Wiley & Sons, Inc., 2013) p. 242.

5. Dan E. Dickinson and David Wood, "Alaska Tax Reform: Intent Met with Oil," (Part 1 of 2), *Oil & Gas Journal*, May 25, 2009, 20–24; Dan E. Dickinson and David Wood, "Alaska Tax Reform: Gas Raises Questions" (Part 2 of 2, *Oil & Gas Journal*, June 1, 2009, 20–26.

6. David A. Wood, "Long Term Fiscal Contractual Stability Proves Elusive, Part 1," *Petroleum Review*, February 2005, 38–42; David A. Wood, "Long Term Fiscal Contractual Stability Proves Elusive, Part 2," *Petroleum Review*, April 2005, 44–48.

7. BP, BP Statistical Review of World Energy, June 2015, page 9, https://www.bp.com/content/dam/bp/pdf/energy-economics/statistical-review-2015/bp-statistical-review-of-world-energy-2015-full-report.pdf

8. BP, BP Statistical Review of World Energy, June 2015, page 8, https://www.bp.com/content/dam/bp/pdf/energy-economics/statistical-review-2015/bp-statistical-review-of-world-energy-2015-full-report.pdf

9. Ibid.

10. International Energy Agency, World Energy Outlook 2015, page 3, https://www.iea.org/Textbase/npsum/WEO2015SUM.pdf

11. Paula Dittrick, "OPEC World Oil Outlook Sees $95/barrel Oil in 2040," *Oil & Gas Journal*, December 28, 2015, http://www.ogj.com/articles/2015/12/opec-world-oil-outlook-sees-95-bbl-oil-in-2040.html

12. Barclays, "E&P Capex: First Double Dip Since '86, January 14, 2016," http://www.oilandgas360.com/ep-capex-first-double-dip-since-86/

13. Ibid.

14. Oil & Gas Journal, "Barclays: Global E&P spending to trend up 5% in 2017," September 7, 2016, http://www.ogj.com/articles/2016/09/barclays-global-e-p-spending-to-trend-up-5-in-2017.html

15. International Energy Agency, World Energy Outlook 2015, p. 3, https://www.iea.org/Textbase/npsum/WEO2015SUM.pdf

Chapter 1

Institutions and Organizations in Theory

DEFINING RISK

In the upstream sector (exploration and production) of the petroleum industry, risk can be divided into two major categories: below-ground risk and above-ground risk. Below-ground risk includes oil and gas resource, technical and environmental uncertainties. Resource uncertainty refers to the presence or absence of oil and natural gas in a specific drilling prospect. This uncertainty takes two forms, a discrete uncertainty and a continuous uncertainty. The discrete uncertainty is related to the risk that there is or is not oil or gas at that location. The continuous uncertainty is related to the amount of oil and natural gas that is recoverable, if a discovery is made. Technical uncertainty refers to the potential, but unknown engineering problems that may be encountered in producing the oil or natural gas. Environmental uncertainty refers to the potential for accidental damage to land or marine life. Below-ground risk is frequently understood to mean the possibility of failure or loss, but it is better understood as a dispersion or range of the possible outcomes. Below ground-risk is therefore a measure of uncertainty or variability of outcome and does not necessarily reflect a high probability of failure or loss.[1]

Above ground risk includes adverse changes in economic conditions (intentional or unintentional), the inability to enforce the terms of a contract, changed fiscal terms (division of revenue) imposed by the host government, resource nationalism, discriminatory regulatory behavior and the placement of geopolitical considerations (national security, energy security and political alliances) above legal and financial commitments. All these contingencies comprise political risk. Political risk can be broadly defined "as the impact of politics on markets. Political risk is influenced by the passage of laws, the foibles of leaders, and the rise of popular movements—in short, all the factors that might politically stabilize or de-stabilize a country."[2] In practical terms, political risk arises from the rent-seeking nature of host governments even in the absence of a regime change.

FOREIGN DIRECT INVESTMENT AND PORTFOLIO INVESTMENT

Lipson noted that "until recently. . . portfolio investments [debt] and direct investment [equity] had long shared roughly similar treatment [levels of protection]. They could be characterized jointly in a single regime governing all foreign capital."[3] He traced the divergence to the late 1960s, when expropriations proliferated, but debt renunciation (and default more generally) did not. He concluded that "*The fundamental difference is the better institutionalization of international finance* [italics in the original]. . . . The answer lies in two basic differences between the network of international lending and equity investments [foreign direct investment]."[4] First, multinational corporations have mostly independent interests in expropriation disputes, making the disputes between two groups of investors and the host government separable issues, but financial organizations have joint interests because of the network of interbank liabilities between them. Second, because the interests of direct investors are separable, their home governments have less incentive to become involved in one another's investment disputes.[5]

ORGANIZATIONS AND INSTITUTIONS

In its classical economic formulation, markets are assumed to be efficient even in the absence of regulation. In its neoclassical formulation, markets are thought to work efficiently most of the time, but may require minimal regulation to maintain an orderly market. The basic assumptions of classical and neoclassical economics are: (1) people have rational preferences among outcomes that can be identified and assigned a value (2) individuals maximize their utility and firms maximize their profits, (3) people act independently on the basis of full and relevant information,[6] and (4) the classical and neoclassical economic models assume that capital is optimally allocated through the action of markets and prices.

Another approach to political economy and economics in general is the "new institutional economics." New institutional economics (NIE) has its origin in two articles written by Ronald Coase, *The Nature of the Firm* and *The Problem of Social Cost*. Institutional economists work within a modified neoclassical framework, which includes consideration of efficiency, that is, transaction and distribution costs.[7]

The NIE approach differs from neoclassical economics in several ways. First, it rejects the assumption that people act independently on full and relevant information. Institutional economists argue that decisions are made on less than complete and accurate information, because the acquisition of information involves costs in time and money and both are constraints. Second, engaging in transactions involves the incurrence of transaction costs and transaction risk, which influence the decision-making process and therefore the decision. Third, the two preceding points make a formal system of rules desirable because formal rules lower information and transaction costs, and reduce the risk associated with specific transactions. The issues addressed by institutional economics include: organizational arrangements, property rights,[8] transaction costs,[9] credible commitments, decisive perceptions, gained control, enforcement mechanisms, modes of governance, persuasive abilities, social norms, ideological values, asset specificity, strategic behavior, bounded rationality, opportunism, adverse selection, moral hazard, human assets, social capital, asymmetric information, incentives to collude, hierarchical structures, bargaining strength, contractual safeguards, surrounding uncertainty and monitoring costs.

The two principal classes of participants in the institutional economic framework are institutions and organizations. There are no universally accepted definitions of the terms "institution" and "organization," but most scholars working within the NIE framework follow Douglass North's definitions. *Institutions* are the "rules of the game," consisting of both the formal legal rules and the informal social norms that govern individual behavior and structure social interactions.

> Institutions are the humanly devised constraints that structure political economic and social interaction. They consist of both informal constraints (sanctions, taboos, customs, traditions, and codes of conduct), and formal rules (constitutions, laws, property rights). Throughout history, institutions have been devised by human beings to create order and reduce uncertainty in exchange. . . . They evolve incrementally, connecting the past with the present and the future; history in consequence is largely a story of institutional evolution in which the historical performance of economies can only be understood as a part of a sequential story.[10]

Organizations are those groups of people and the governance arrangements they create to coordinate their group action against other groups also performing as organizations, for example, business firms, universities, clubs, medical associations and unions.[11] These definitions are used throughout this book.

Coase, North, Williamson and others argue that the role of institutions has been underestimated and that institutions are not a neutral or unchanging background against which rational individuals and organizations make decisions. Rather institutions "together with the standard constraints of economics . . . define the choice set and therefore determine transaction and production costs and hence the profitability and feasibility of engaging in economic activity."[12] Institutions not only define the "choice set," but also influence organizations, consequently changing both the organizations and the institutions over time. "Incremental change comes from the perceptions of the entrepreneurs in political and economic organizations that they could do better by altering the existing institutional framework at some margin."[13]

Institutions (contracts, domestic laws, treaties, international law, formal rules and informal customs) and the entities that implement them (regulatory agencies, financial exchanges, nongovernmental agencies and intergovernmental agencies) are involved in virtually every stage of the oil and gas exploration, development and production process. This involvement extends beyond the neoclassical assumptions regarding rules and regulations needed to maintain an orderly market and includes laws regarding capital formation, ownership structure, foreign direct investment, capital allocation, domestic and international product pricing, taxation, capital repatriation and social service expenditures.

CENTRAL QUESTIONS

What institutions and strategies are available for managing political and investment risk in the international oil and gas industry? How and when did they develop? How are they used by organizations in the oil and gas industry? How effective have they been?

The institutions that facilitate the management of risk include oil and gas exploration contracts, domestic courts, national constitutions, bilateral investment treaties, multilateral investment treaties, governmental and nongovernmental regulatory agencies and international energy forums. The strategies for managing risk include corporate finance, joint ventures, project finance, alliances and energy diplomacy. The organizations that manage risk include international oil companies, oil service companies, national oil companies and public and private providers of financial capital and political risk insurance.

The evidence in this book is consistent with North's conclusion that "incremental change [in institutions] comes from the perceptions of the entrepreneurs in political and economic organizations that they [the entrepreneurs] could do better by altering the existing institutional framework at some margin"[14] and that "institutions do not emerge spontaneously to create and nurture the market, but rather reflect the interests of those players in a position to put them in place."[15] However, the process by which these institutions are created and evolve is easier to describe in theory than it is to explain in practice, because institutional development takes place in steps; it involves several organizations, some trying to change the existing institutional structure and others trying to preserve it.

The institutions in the international oil and gas industry can be arranged in a hierarchy based on their relative importance. The most important institutional mechanisms used by international oil and natural gas companies are contracts, informal relationships and transparency. The second most important institutions include domestic law, bilateral treaties and international courts and tribunals. A third line of defense includes multilateral treaties and international forums. Only a small number of these institutions are the direct result of developments in the oil and gas industry. Most have developed in response to the needs of foreign investors in general and have then been adapted for use in the oil and gas industry.

NOTES

1. David Wood, *Petroleum Economics, Risk and Opportunity Analysis*, Chapter 10 in Betty J. Simkins and Russell E. Simkins, Energy Finance and Economics—Analysis and Valuation, Risk Management, and the Future of Energy (Hoboken, NJ: John Wiley & Sons, Inc., 2013), p. 239.
2. Ian Bremmer, "Managing Risk in an Unstable World," *Harvard Business Review*, June (2005).
3. Charles Lipson, *Standing Guard – Protecting Foreign Capital in the Nineteenth and Twentieth Centuries* (Berkeley: University of California Press, 1985), p. 111.
4. Ibid.
5. Ibid., p. 174–175.
6. E. Roy Weintraub. *The Concise Encyclopedia of Economics*, "*Neoclassical Economics*." http://www.econlib.org/library/Enc1/NeoclassicalEconomics.html.
7. Oliver E. Williamson, *Markets and Hierarchies, Analysis and Antitrust Implications: A Study in the Economics of Internal Organization* (Free Press, 1983).
8. Dean Lueck, *The New Palgrave Dictionary of Economics*, 2nd Edition, 2008, "Property Law and Economics."
9. M. Klaes, *The New Palgrave Dictionary of Economics*, 2nd Edition. Abstract. "History of Transaction Costs." ed. Steven N. Durlauf and Lawrence E. Blume. Palgrave Macmillan, (June 13, 2008). http://www.dictionaryofeconomics.com/article?id=pde2008_T000239&edition=current&q=transaction%20costs&topicid=&result_number=1
10. Douglass C. North, "Institutions," *Journal of Economic Perspectives*, Vol. 5, No. 1, Winter (1991), p. 97–112.

11. Ibid., p. 97.
12. Ibid., p. 97.
13. Douglass C. North, *Institutions, Institutional Change and Economic Performance* (Cambridge, UK: Cambridge University Press, 1990), p. 8.
14. Ibid., p. 8.
15. Ibid., p. 89–90.

Chapter 2

Organizations in Practice

The organizations in the international oil and gas industry can be divided into three broad categories: (1) national governments, (2) NOCs and (3) IOCs. This chapter examines the politics, policies and objectives of each of these types of organizations.

NATIONAL GOVERNMENTS

National governments are the representatives of the people and therefore, at least in theory, act to enhance the welfare of the people in that country. In countries exporting oil and natural gas, this principally consists of the development of energy policies related to the discovery, production and sale of oil and gas. Energy policy includes approving the organizations that will be granted drilling rights in the country, the rate of production allowed, and the division of revenue between the national government and the IOCs or foreign NOCs operating in the country. The energy policy of national governments can change over time and this represents in the broadest sense, the risk to foreign direct investment. This section examines the politics, policies and objectives of some of the larger oil exporting countries and their governments.

Saudi Arabia

Saudi Arabia possesses 15.7 percent of the world's proved oil reserves and was the largest producer of petroleum liquids in the world in 2015. In 2015, Saudi Arabia produced an average of 12.104 MBPD[1] and exported an estimated 7.163 MBPD of crude oil.[2] In 2015, Saudi Arabia exported an average of 1.051 MBPD of crude oil to the United States, which accounted for 14.3 percent of total U.S. crude oil imports.[3] Saudi Arabia's economy remains heavily dependent on petroleum exports. In 2015, petroleum exports accounted for almost 76.9 percent of Saudi Arabia's total export revenues.[4]

The state oil company, Saudi Arabian Oil Company, has exclusive rights to explore and produce oil and natural gas in the country. Saudi Arabia's Ministry of Petroleum and Mineral Resources and the Supreme Council for Petroleum and Minerals have oversight of the Saudi Arabian Oil Company and the oil and gas industry. The Supreme Council is composed of members of the royal family, industry leaders and government ministers and is responsible for petroleum and gas policy, including contract review and long-term strategic planning. Saudi Arabian Oil Company's financial performance is in the upper middle level in the classification of performance, one level below Statoil and Petroleo di Brasiliero (Petrobras), which are considered the two best managed NOCs, and one level above PDVSA and PEMEX.[5]

Viewed from the outside, internal politics does not appear to be an obstacle to the formation and implementation of a consistent energy policy in Saudi Arabia. In addition, Saudi Arabia's policies and production level have been consistent with its public statements, which is to provide enough oil to maintain the world price of oil at a level that will not lead to a global recession. However, the spot price of WTI crude oil declined from $106 per barrel in June 2014 to $29.71 per barrel on February 8, 2016. During this same period, Saudi Arabian oil production increased from 11.393 MBPD in 2013, to 11.505 MBPD in 2014 and 12.014 in 2015.[6] Saudi Arabia's decision to continue to increase production during a period in which the supply of crude oil exceeded demand and the price of crude oil declined significantly, suggests that Saudi oil production was being driven by something other than price stability or the avoidance of a global recession. On September 10, 2016, the price of WTI closed at $45.74 per barrel.

Given the significant reduction in the price of crude oil, Saudi Arabia might have been expected to reduce its oil production to bring the global supply of oil into balance with demand. However, it continued to increase production. There are several possible reasons for this. First, Saudi Arabia may not have wanted to reduce its market share of crude oil exports relative to other countries, particularly those countries in Organization of Petroleum Exporting Countries (OPEC). Second, Saudi Arabia's marginal cost of oil production is significantly lower than other countries within and outside of OPEC. They can therefore operate at a profit even at substantially lower oil prices. Third, they may want to discourage the production of unconventional oil and gas (shale oil and shale gas) in the United States and Canada. Fourth, they may want to keep economic and financial pressure on the Shia government of Iran, which from time to time has sought hegemony in the Middle East and attempted to destabilize the Sunni-led regimes in Saudi Arabia, Bahrain and Kuwait.

Russia

In 2015, Russia produced an average of 10.980 MBPD[7] of petroleum liquids and exported an estimated 5.291 MBPD,[8] making it the third-largest producer of crude oil (after Saudi Arabia and the United States). Russia was the second-largest producer of natural gas in 2015, with 573.3 billion cubic meters (after the United States).[9] In 2015, the state-owned company *Rosneft* produced 2.602 MBPD, 23.7 percent of Russia's oil production of 10.980 MBPD. LUKoil, a privately owned company, produced 1.779 MBPD, 16.2 percent of Russia's oil production.

Russia produced 573.3 cubic meters of natural gas in 2015. The state-owned company *Gazprom* dominates Russia's upstream gas industry, producing approximately 73 percent of Russia's total natural gas output. *Gazprom* also controls most of Russia's natural gas reserves. More than 65 percent of proved gas reserves are directly controlled by the company and additional reserves are controlled by *Gazprom* through joint ventures with other companies.[10] Russia depends on oil and natural gas export revenue to fund approximately 50 percent of the federal budget.[11]

After the dissolution of the Soviet Union in 1991, Russia privatized its oil industry by selling state-owned companies to private investors. Beginning in the late 1990s, a small number of private sector companies led the growth in the Russian oil industry. Although foreign companies can invest in Russia, the investment is generally made in partnership with a Russian company, usually *Rosneft* if the investment is in the oil sector and Gazprom, if the investment is in the gas sector.[12]

The Russian oil and gas industry is regulated by the Ministry of Natural Resources which issues field licenses, monitors compliance with license agreements and levies fines for violations of environmental regulations. The Ministry of Finance implements and enforces tax policy and the Ministry of Economic Development sets and collects tariffs. The Ministry of Energy supervises energy policy formulation and enforcement. Within these ministries, there are agencies with more specific regulatory responsibility. These include the Federal Energy Commission, which sets oil transportation tariffs, the Commission for State Policy on the Oil Market, which establishes policy for regulating crude oil and refined product markets, and the Commission on Protective Measures in Foreign Trade & Customs and Tariff Policy, which sets crude oil export tariffs.[13]

Since 1999, the Russian oil and gas industry has undergone renationalization. Russia's tax and environmental regulatory agencies have been instrumental in helping to affect this renationalization. This is discussed in more detail in the TNK-BP case study in chapter 6. Russia's energy policy is directed at three principal objectives. The first is to use oil and gas export revenue to finance the Russian federal budget. The second is to use oil and gas exports to create and maintain political alliances with countries bordering Russia. The third is to develop the domestic economy; however, this third objective has received less attention from the Russian government than the first two objectives. A combination of economic sanctions were imposed on Russia by the United States and its coalition partners in response to Russia's annexation of Crimea; and its incitement of and assistance to Russian separatists in Eastern Ukraine. The decline in the spot price of crude oil WTI from $106 per barrel in June 2014 to $29.71 per barrel on February 8, 2016 and the country's heavy reliance on oil and gas exports for foreign exchange reserves led to a run on the ruble and a recession in Russia in 2015 and 2016. The Russian GDP decreased by 3.7 percent in 2015 and the IMF expected the Russian economy to decrease by another 1.2 percent in 2016.[14]

Venezuela

In 2015, Venezuela produced an average of 2.626 MBPD[15] and exported an estimated 1.974 MBPD of petroleum liquids.[16] PDVSA is the largest employer in Venezuela. Oil production accounts for 25 percent of the country's GDP and 95 percent of its export earnings.[17]

In the 1990s, Venezuela adopted more market-oriented policies in its oil and gas sector. In 1999, Venezuela passed the Gas Hydrocarbons Law, to diversify the Venezuelan economy by promoting gas development and expanding the role of natural gas in Venezuela's energy sector. This legislation allowed private operators to own 100 percent of gas projects, in contrast to the ownership rules in the oil sector which required that Venezuela own a majority interest in all new oil projects. The Gas Hydrocarbons Law also reduced royalty and income tax rates on gas projects relative to the rates applied to oil projects.[18]

Hugo Chavez was elected president of Venezuela in 1999. After his election, the Venezuelan government began to increase its participation in the Venezuelan oil industry.[19] In 2002, almost half of PDVSA's employees protested the policies of President Chavez and his administration by going on strike. After the strike, PDVSA fired 18,000 employees and restructured the organization to strengthen the government's control over the company and increase the loyalty of middle and upper management. In 2006, President Chavez began the nationalization of oil exploration and production in Venezuela, by mandating the renegotiation of ownership of all projects and giving PDVSA a minimum of 60 percent ownership in all oil and gas projects (old and new).[20]

In 2007, President Chavez introduced a public referendum on several proposed constitutional amendments, one of which would have allowed the Venezuelan government a controlling interest in all new natural gas projects like the constitutional law already governing the Venezuelan oil sector. However, Venezuelan voters defeated the referendum in December 2007.[21] Chavez's successor, Nicolas Maduro, was elected president of Venezuela on April 15, 2013, and has continued the policies of his predecessor. Crude oil production in Venezuela has decreased from 3.336 MBPD in 2006 to 2.626 MBPD in 2015.[22]

Mexico

In 2004, Mexico produced an average of 3.848 MBPD and exported an estimated 1.792 MBPD of petroleum liquids. In 2015, Mexico produced an average of 2.558 MBPD and exported an estimated 1.14 MBPD of petroleum liquids.[23] In early December 2013, after years of government ownership of the oil and gas sector and declining production, Mexico's president Enrique Pena Nieto proposed significant energy reforms. On December 12, 2013, the Mexican Senate passed a new law, with ninety-five votes in favor and twenty-eight against, allowing domestic and foreign companies to explore and develop oil fields, for the first time since 1938. This legislation was subsequently referred to the lower Mexican Chamber of Deputies where it was approved by a vote of 345 in favor and 134 against. President Pena Nieto signed the energy law on December 20, 2013.[24]

Advocates of the new law argued the status quo was unacceptable, because Mexico was facing depleted reserves and had already experienced a decade of declining crude oil output.[25] The law is intended to restore the efficiency of the oil and natural gas industry in Mexico, which is suffering from corruption, inefficiency and an inability to successfully implement new technologies. The reforms also propose improving the transparency and effectiveness of national finance.[26]

The reforms would allow exploration and production contracts that include licenses, production-sharing agreements (PSAs), profit-sharing agreements and service contracts. Prior to this legislation only service contracts, in which foreign companies were paid for services rendered, were allowed. Under the new legislation, PEMEX, will remain a state-owned company, but will be given more budgetary and administrative autonomy and must bid competitively with other oil and gas companies on new projects.

On June 19, 2014, the conservative National Action Party said that it would only support the energy reforms if President Pena Nieto agreed to electoral reforms that could have the effect of weakening his party's (Institutional Revolutionary Party, PRI) hold on power.[27] In addition, Mexico's left-wing parties had previously announced that they intended to promote a 2015 referendum to repeal the new energy laws already passed.[28] However, on August 6, 2014, the Mexican Senate voted 78–26 in favor of a package of legislation that implements these changes to the Mexican oil industry and President Pena Nieto signed the laws on August 11, 2014. On October 1, 2014, Mexico's state oil company PEMEX and Exxon Mobil Corp signed a noncommercial agreement to jointly explore potential upstream and downstream business opportunities.[29] On the same day, formation of a new private sector Mexican oil company, named Sierra Oil & Gas, was announced. Sierra Oil & Gas has secured equity investment commitments of $525 million from U.S. private equity firms Riverstone Holdings LLC and EnCap Investments.[30] In 2015, the round one auction offered fourteen offshore blocks, but only two bids were deemed adequate and awarded. However, the pace and success of the auctions have improved. By August of 2016, Mexico had completed three rounds of auctions on oil fields and thirty production contracts had been awarded.[31]

Developments in Mexico illustrate several of the features of the oil and natural gas industry described earlier: (1) the depletion of "easy oil reserves," (2) the need to employ more advanced technology in oil-producing regions where the "easy oil" has been depleted and (3) the uncertainty created by a continuing struggle between populist nationalism and resource sovereignty on the one hand and economic efficiency and investor protection on the other.

China

In 2015, China produced 4.309 MBPD of oil making it the fifth-largest oil producer in the world, after the United States (12.704 MBPD), Saudi Arabia (12.014 MBPD), Russia (10.980 MBPD) and Canada (4.385 MBPD). In 2015, China consumed 11.968 MBPD, making it the second-largest oil consumer in the world, after the United States (19.396 MBPD).[32] China's energy policies are driven by its increasing demand for oil and its dependence on oil imports. The National Development and Reform Commission (NDRC), a department within China's State Council, is the principal policy making, planning and regulatory authority for the Chinese energy industry. Four other ministries also oversee various aspects of country's oil policy.[33]

In July 2008, the government created the National Energy Administration (NEA) to serve as the principal energy regulator. The NEA and the NDRC are responsible for approving new energy projects in China, setting domestic wholesale energy prices and implementing the central government's energy policies.[34] In January 2010, the government formed a National Energy Commission to consolidate and unify energy policies among the various agencies under the State Council.[35] China's NOCs have acquired increasing influence in the formation of energy policies in China.

Between 1994 and 1998, the Chinese government reorganized most state-owned oil and natural gas assets into two vertically integrated firms that owned both upstream and downstream assets, the China National Petroleum Corporation (CNPC) and the China Petroleum and Chemical Corporation (SINOPEC). These two companies operate a wide range of local subsidiaries, and together control most of China's upstream and downstream oil markets. CNPC is the largest upstream oil company in China and accounts for an estimated 53 percent and 75 percent of China's total crude oil and natural gas production, respectively. (This includes CNPC's publicly traded subsidiary PetroChina.) Sinopec focuses on refining and distribution, which account for 76 percent of Sinopec's revenues. Other state-owned oil firms have emerged over the past several years. For example, the China National Offshore Oil Corporation (CNOOC) is responsible for offshore oil exploration and production.[36]

Onshore oil exploration and production in China is limited to China's NOCs, however, IOCs have participated in the exploration of offshore blocks and prospects that require more advanced exploration and production technology. These projects are carried out primarily through production-sharing contracts (PSCs) and joint ventures (JVs) with IOCs. Chinese investment laws require that China's NOCs must hold the majority interest in each PSA and retain the right to become the operator after development costs have been recovered by the partners in the PSA.[37]

The Communist Party of China (CPC) is the sole governing party in China, but it coexists with eight other legal parties that constitute the United Front. It is therefore more appropriate to discuss China's internal politics and policy in the context of competing bureaucracies rather than political parties. This competition includes disputes regarding internal pricing and subsidies, greater state sponsorship of one energy source over another, conflict over the priority to be given to environmental protection and national economic development, and the extent to which state-owned companies (CNPC, SINOPEC and CNOOC) are permitted to negotiate their own deals with IOCs and host states.

Nigeria

In 2015, Nigeria produced 2.352 MBPD of petroleum liquids,[38] making it the thirteenth-largest producer in the world, and exported 2.114 MBPD.[39] The Nigerian National Petroleum Corporation (NNPC) was created in 1977 to oversee the Nigerian oil and natural gas industry and promote oil and natural gas development. In 1988, the NNPC was divided into twelve subsidiary companies, each with the responsibility to regulate a specific subsector of the Nigerian oil and gas industry. The Department of Petroleum Resources (DPR), a department within the Ministry of Petroleum Resources, is also an important regulator of the Nigerian oil and gas industry.[40]

Most of Nigeria's major oil and gas projects are funded through JVs between IOCs and NNPC, in which NNPC is the majority shareholder. PSAs are usually employed for deepwater projects and generally include more attractive terms for the IOC than the projects onshore or in shallow water, to provide an incentive for the IOCs to invest in deepwater projects.[41]

The Petroleum Industry Bill (PIB) was initially proposed in 2008 and was expected to change the structure and fiscal terms governing the oil and natural gas sectors in Nigeria. The bill was intended to ensure that the management and allocation of petroleum revenues in Nigeria were done in a manner consistent with principles of good governance, transparency and sustainable development. The most recent version of the PIB was submitted to the Nigerian National Assembly on July 18, 2012.[42] As of September 6, 2016, the passage of the PIB was again delayed because of the continuing armed conflict in the Niger Delta region, which is Nigeria's main oil export terminal.[43]

Some of the most contentious provisions of the PIB are the potential renegotiation of existing contracts with IOCs, changes in tax and royalty structures, deregulation of the downstream sector, restructuring of NNPC, a concentration of oversight authority in the Ministry of Petroleum Resources and a mandatory contribution by IOCs of 10 percent of monthly net profits to the Petroleum Host Communities Fund. The PIB could make some projects financially unattractive, particularly deepwater projects.[44]

Because of the regulatory uncertainty created by the delay in passing the PIB, there has not been a licensing round for oil exploration in Nigeria since 2007. This regulatory uncertainty extends to the natural gas industry as well. The result is that less investment has been made in new projects than might otherwise have been made. Developments in Nigeria illustrate several of the features of the international oil and natural gas industry described earlier: (1) the depletion of "easy oil reserves," (2) the need to employ more advanced technology and (3) the uncertainty created by a continuing debate over changes in national energy laws.

Iraq

Iraq's oil production has increased from 3.141 MBPD in 2013, to 3.285 MBPD in 2014 and 4.031 MBPD in 2015.[45] The majority of Iraqi oil production comes from just three very large oil fields: (1) Kirkuk, (2) the North Rumaila field in southern Iraq and (3) the South Rumaila field, also in southern Iraq, of which approximately 75 percent comes from these two southern fields.

The Iraqi Ministry of Oil provides oversight of oil and gas exploration and production in all areas of Iraq except the Kurdish territory through its operating entities, the North Oil Company and the Midland Oil Company in the north and central regions of Iraq; and the South Oil Company and the Missan

Oil Company in the southern region of Iraq. Oil production in the northern region is controlled by the Kurdistan Regional Government (KRG). Production in the region controlled by the KRG is estimated to be .450 MBPD.[46]

Despite the violence created by The Islamic State in Iraq and Syria (ISIS) also known as The Islamic State in Iraq and the Levant, oil production in Iraq has continued to increase. ISIS's activities have not affected oil production in Iraq because most of Iraq's oil production is in the southern part of the country. In addition, the oilfields controlled by the KRG in northern Iraq have not been significantly affected either, even though the fighting between ISIS and the Iraqi army came very close to those oilfields.[47]

The proposed Hydrocarbon Law has been under review in the Council of Ministers since October 26, 2008. The law would govern Iraq's contracts with and regulation of oil and gas companies, but it had not received final passage as of September 11, 2016. The insistence on centralizing government policy making in the Shia-dominated government in Baghdad by the ruling Dawa Party under the leadership of former Iraqi prime minister Nouri al-Maliki only exacerbated the conflict between the Sunni's, Shia's and Kurds in Iraq. The election of Haider al-Abadi in 2014, who appears be a more inclusive leader, may help bring Sunnis, Shias and Kurds closer together on this and other issues, making cooperation on oil production and oil policy possible.

NATIONAL OIL COMPANIES

NOCs control a substantial part of world oil production and world oil reserves. In 2014, NOCs were estimated to control 75 percent of world proved oil reserves and 58 percent of world oil production.[48] In 2014, the eight largest NOCs measured by daily oil production were Saudi Arabian Oil Company (9.5 MBPD), Rosneft (4.1 MBPD), Iranian National Oil Company (3.4 MBPD), China National Petroleum Company (3.3 MBPD), Kuwait Petroleum Corporation (2.8 MBPD), Petroleos de Venezuela (2.85 MBPD), Abu Dhabi National Oil Company (2.7 MBPD) and Nigerian National Oil Company (2.4 MBPD).[49] Consequently, they play a critical role in the balance between supply and demand. Understanding the policies and behavior of NOCs is therefore important for understanding the operation of the international oil and gas industry.

Victor identified four approaches to understanding NOCs: (1) theoretical and historical studies to explain why NOCs exist, (2) operational studies of NOCs which have attempted to explain their wide variation in financial and economic performance, (3) organizational and political studies that have tried to explain the abnormal politics in resource-rich countries and (4) studies advocating economic and political reform of NOCs, particularly those

NOCs that have performed poorly or engaged in perverse political behavior or both.[50] The next two sections examine questions 1 and 3.

Why NOCs Exist

Tordo et al prepared a study that examined the question "Why do NOCs exist?"[51] Using that study, Victor identified four reasons why governments have created and maintained NOCs rather than authorize the creation of private sector companies and simply impose taxes on them.[52]

The first of the four reasons is the weakness of public institutions. National governments must perform many functions, including the regulation of a complex oil and gas industry. The institutions needed to regulate the industry are often weak or nonexistent in developing countries. Governments may create an NOC to improve the collection of industry information and to facilitate the regulation of the oil and gas industry by giving the government direct access to the country's oil and gas operations through the NOC.[53, 54] A single NOC can track and allocate revenue streams more easily and therefore provide fiscal oversight of the country's oil and gas industry. For example, in Angola one of the main functions of its NOC, Sonangol, is to regulate the oil and gas sector and oversee the behavior of the large number of IOCs operating in the country.[55]

The second reason is the inadequacy of the domestic private sector. The suppliers and stages of production needed to explore for and produce oil and gas often do not exist in a country with previously unknown hydrocarbon resources or one that is still developing; furthermore maximizing the value of the country's oil and gas resources requires coordination among many suppliers and stages of production. Governments often create an NOC to help encourage the orderly development of these suppliers and stages of production in the country's oil and gas industry.[56] For example, many NOCs are given the responsibility for the development of domestic industry suppliers (e.g., drill pipe, drilling fluids, seismic exploration and waste water treatment) and customers (utilities, industrial and automotive). For example, in Saudi Arabia, the Saudi Arabian Oil Company led the development of the nation's petrochemical industry.[57, 58]

A third reason is the desire for internal control over the revenue created by oil and gas production. Control over the cash inflow from the production and sale of oil and gas is important to a host government. This is important, whether it intends to use the revenue for public purposes or is intent on diverting some or all of it for private gain.[59, 60] In countries whose governments primarily depend on the revenue derived from the export of oil and gas to finance their national budgets and in which *rentier* politics predominates, one way to control the rents is to control the NOC. This explanation is usually reformulated in public statements as a matter of national pride and the people's desire for the state to exercise sovereignty over the country's oil resources.[61]

The fourth reason is *the desire to project economic and political power abroad.* An NOC can be used to exert political influence abroad as well as at home; for example, Russia has at various times used the threat of interruption or termination of its supply of gas to Western Europe and the countries that were previously a part of the Soviet Union to retain influence in these countries.[62] These four explanations are related and in most cases, more than one is operating in a country at any given time.

The Political Behavior of NOCs

The decision to create an NOC is a political act, and because of their size and importance to the national economy, NOCs inevitably become involved in national politics. The studies of political behavior of NOCs can be divided into two categories. One category emphasizes the principal-agent relationship between the host government and the NOC, in which the principal (the state and its people) try to encourage an agent (the NOC) to align its behavior with the principal's interests.[63]

Another group of studies demonstrate how NOCs become political actors to advance their own agenda and increase their power. For example, Mommer[64] demonstrated how NOCs that began as rent collectors for the state (e.g., PDVSA) acquired a larger range of functions over time, including oil production and refining operations abroad.[65] A large number of case studies have shown how NOCs have become states within a state, insulating themselves from review by the government that created them and so creating the freedom to pursue their own interests. Russia's *Gazprom* is one of the more obvious examples.[66] Some of the literature on Saudi Aramco points to similar conclusions.[67] The next section examines the history, politics, policies and objectives of IOCs.

INTERNATIONAL OIL COMPANIES

A Brief History

In 1951, the seven major IOCs (Exxon, Shell, British Petroleum, Gulf, Texaco, Mobil and Chevron) controlled 98 percent of the world oil trade.[68] In the 1970s, IOCs still held approximately 85 percent the world's known hydrocarbon reserves.[69] The IOCs still marketed 75.4 percent of all internationally traded oil in 1976, however, between 1970 and 1980, a substantial portion of the world's oil reserves were nationalized by the governments of the oil-producing countries. For example, in 1976 the nationalization of the Kuwait Oil Company was completed and is 100 percent owned by the government of Kuwait.[70] In 1979, during the Islamic Revolution, the National Iranian Oil Company (NIOC) was nationalized and is 100 percent owned by the government of Iran.[71] In 1974, 60 percent of Saudi Aramco was owned by the Saudi government, however, since 1980 it is 100 percent owned by the Saudi government.[72] The immediate effect of these nationalizations was to substantially reduce the oil and gas reserves under the control of the IOCs.

Consequently, the share of oil and gas sales made by NOCs after the mid-1970s rose from 24.6 percent in 1976 to 42.2 percent in 1979. By 1980, direct marketing of oil by NOCs approached 50 percent of all internationally traded oil.[73] In 2012, it was estimated that NOCs owned a minimum of 73 percent of the world's oil reserves and accounted for 61 percent of oil production; and the NOCs owned 68 percent of the world's natural gas reserves and accounted for 52 percent of natural gas production.[74] In 2014, NOCs were estimated to control 75 percent of world proved oil reserves and 58 percent of world oil production.[75]

The secondary effect of these nationalizations was to call into question the future of IOCs. The IOCS responded by merging. For example, Exxon and Mobil merged in 1999 and Conoco and Phillips Petroleum merged in 2002. The strategic rationale for the IOCs merging included, (1) reducing costs in the merged companies by reducing the duplication of personnel associated with administrative functions (finance, accounting, legal), (2) reducing the number of personnel associated with exploration and production made necessary by the availability of fewer exploration blocks and (3) the belief that the merged companies would be better able to successfully develop very large oil and natural gas projects in difficult environments, often involving complex geology and significant technical risk.[76] Consequently, much of the research on IOCs between 1985 and 2005 focused on their future role in the industry.

However, the IOCs succeeded in making this transition and are still among the largest oil and gas companies in the world.[77] In 2012, ExxonMobil, Royal Dutch Shell, Chevron Corp, BP Plc and ConocoPhillips had a combined net income of $121.2 billion, which was 30.1 percent of all oil and gas companies reported profits in that year. (These figures do not include the countries of the Middle East and Africa which do not report net income). However, in 2015, these five companies reported a combined net income of $12.7 billion.[78] This reduction in net income was the result of substantially lower oil prices in 2015.

In 2012, these five companies produced 7.47 MBPD of oil or 8.4 percent of the 89.8 MBPD produced worldwide in 2012; and 31.7 billion cubic feet per day of natural gas or 7.8 percent of the 404.0 billion cubic feet produced per day, worldwide in 2012.[79] In 2015, these five companies produced 7.42 MBPD of oil and 31.7 billion cubic feet per day of natural gas.[80] (These numbers do include the countries of the Middle East and Africa.) The IOCs therefore still control substantial capital inflows that can be invested in future exploration and production projects.

In 2012, these five companies accounted for $136.8 billion or 23.6 percent of the $579.5 billion spent on CAPEX for oil and natural gas exploration worldwide.[81, 82] (The $579.5 billion does not include the capital expenditures and exploration expense made by the NOCs in the Middle East and Africa and is therefore lower than the $617.2 billion estimated by Barclays Bank for 2012.) Barclays estimated the CAPEX made by the NOCs of the Middle East and Africa at $26.3 billion and $23.8 billion, respectively, in 2012. Adding these two figures to $579.5 billion gives $629.6 billion ($579.5 + $26.3 + $23.8 = $629.6), which is close to the $617.2 billion Barclays estimated for global CAPEX in 2012.)[83] In 2015, these five companies accounted for $115.1[84] billion or 22.9 percent of the $502.0 billion spent on CAPEX for oil and natural gas exploration worldwide in 2015.[85] (The $502 billion does include the CAPEX made by NOCs in the Middle East and Africa.)

However, in 2012, these five firms accounted for only 33.0 billion barrels of proved liquid reserves (oil and LPG) or 2.0 percent of liquid reserves worldwide (conventional and unconventional) of 1,687.3 billion barrels.[86, 87, 88] In 2015, these five firms accounted for only 32.9 billion barrels of proved liquid reserves (oil and LPG)[89] or 1.9 percent of liquid reserves worldwide (conventional and unconventional) of 1,697.6 billion barrels.[90]

Changing Role of the International Oil Companies

Between 1995 and 2005, the role of the IOCs evolved from technically integrated and self-sufficient exploration and development companies to general contractors that coordinate the operations of many suppliers and subcontractors that perform specific tasks (e.g., producing and analyzing seismic data, providing drilling rigs and crews and providing a wide range of oil field services). The IOCs also provide the project management and organizational skills these large projects require. The largest IOCs also function as bankers by providing or arranging the financing needed to explore and develop large fields in difficult environments.[91] For this reason, IOCs still have bargaining leverage at the beginning of negotiations with a host country.[92]

The Politics of International Oil Companies

IOCs are not primarily political actors, but rather economic actors. They tend to be involved in political issues only to the extent that the actions of governments affect the economics of their operations and their future profitability. However, the price and the availability of oil and natural gas are of intense interest to governments and their citizens in developed and developing countries. Oil and natural gas is therefore the object of considerable government legislation and regulation.

Bradley characterizes the political behavior of private sector oil companies as either defensive or offensive. Defensive behavior refers to oil companies trying to block legislation that they believe will increase their costs or reduce their revenues and net income. Offensive behavior refers to oil companies initiating legislation that they believe will reduce their costs and increase their revenues and net income. This latter example is the kind of political behavior that Douglass North has in mind when he says, "Incremental change [in institutions] comes from the perceptions of the entrepreneurs in political and economic organizations, that they could do better by altering the existing institutional framework at some margin."[93]

Gourevitch proposed a more nuanced explanation for policy formation and institutional development. He described five approaches to explaining the selection of a specific policy choice. These are the *production profile*, *intermediate associations*, *state structure*, *economic ideology* and the *international system*. The *production profile* approach emphasizes the preferences of societal actors shaped by the actors' economic or competitive position in the international and domestic economy. The *intermediate associations* emphasizes the large network of associations that lie between the government and business actors and are created to represent business actors and deal with the interactions between them and the government. Since societal forces and representative associations must act through the state structure to attain policy objectives, the *structure of the state*, its rules and institutions, can therefore have a substantial effect on the outcome. Rules shape the process of aggregation, and different rules favor different outcomes. This is the institutional explanation described and advanced by North. The *economic ideology* interpretation of economic policy choices explains outcomes in terms of national traditions and values concerning the economy. The four preceding approaches point to factors internal to the country to explain politics and policy, but it is clear from the material presented in previous sections that the structure of the *international system* has an important influence on the factors that comprise the domestic internal systems.

North does not address economic performance as a function of events arising from outside the country, but rather limits his discussion to issue arising within the country. He only discusses the international environment and the impact of one country on another when he considers the potential for diffusion of economic and political ideas across countries. All five approaches outlined by Gourevitch have been dominant at some point in the political history of the oil and gas industry.[94]

Whether the oil and gas industry's political behavior is driven by offensive or defensive considerations, or by one of the explanations proposed by Gourevitch, the fact remains that the oil and gas industry has a considerable presence in governments around the world. For example, in 2015, the oil and gas industry, including exploration, production, refining, transportation and distribution, spent $129.9 million on behalf of 162 clients and employed 663 lobbyists in Washington, DC.[95]

NOTES

1. BP, *BP Statistical Review of World Energy June 2016*, http://www.bp.com/content/dam/bp/pdf/energy-economics/statistical-review-2016/bp-statistical-review-of-world-energy-2016-full-report.pdf, last accessed on September 10.
2. Organization of Petroleum Exporting Countries (OPEC), http://www.opec.org/opec_web/en/about_us/169.htm.
3. U.S. Energy Information Administration, https://www.eia.gov/dnav/pet/pet_move_impcus_a2_nus_epc0_im0_mbblpd_a.htm, last accessed on September 10, 2016.
4. World's Top Exports, *Saudi Arabia's Top Ten Exports*, April 1, 2016, http://www.worldstopexports.com/saudi-arabias-top-10-exports/.
5. David G. Victor, David R. Hults and Mark Thurber, *Oil and Governance – State-owned Enterprises and the World Energy Supply.* (Cambridge: Cambridge University Press, 2012), p. 898.
6. BP, *BP Statistical Review of World Energy June 2016*, http://www.bp.com/content/dam/bp/pdf/energy-economics/statistical-review-2016/bp-statistical-review-of-world-energy-2016-full-report.pdf, last accessed on September 10, 2016.
7. Ibid.
8. Steven Bierman and Dina Krennikova, Bloomberg, "Russian Exports set for Record as Europe Competition Grows," July 3, 2016, http://www.bloomberg.com/news/articles/2016-07-03/russian-oil-exports-set-for-record-as-europe-competition-grows.
9. BP, *BP Statistical Review of World Energy June 2016*, http://www.bp.com/content/dam/bp/pdf/energy-economics/statistical-review-2016/bp-statistical-review-of-world-energy-2016-full-report.pdf, last accessed on September 10, 2016.
10. U.S. Energy Information Administration, *Russia*, EIA, November 26, 2013, p. 6–7.
11. U.S. Energy Information Administration, *Russia*, EIA, July 28, 2015, p. 1.
12. U.S. Energy Information Administration, *Russia*, EIA, November 26, 2013, p. 10.
13. Ibid., p. 6–7.
14. CNN Money, "Russia has just marked 18 months of a deep and painful recession," August 11, 2016, http://money.cnn.com/2016/08/11/news/economy/russia-economy-recession-six-quarters/.
15. BP, *BP Statistical Review of World Energy June 2016*, http://www.bp.com/content/dam/bp/pdf/energy-economics/statistical-review-2016/bp-statistical-review-of-world-energy-2016-full-report.pdf, last accessed on September 10, 2016.

16. Organization of Petroleum Exporting Countries (OPEC), *Venezuela Facts and Figures*, http://www.opec.org/opec_web/en/about_us/171.htm, last accessed on September 11, 2016.
17. Ibid.
18. U.S. Energy Information Administration, *Venezuela*, EIA, October 2012, p. 3.
19. Ibid., p. 3.
20. Ibid.
21. Ibid., p. 9.
22. BP, *BP Statistical Review of World Energy June 2016*, http://www.bp.com/content/dam/bp/pdf/energy-economics/statistical-review-2016/bp-statistical-review-of-world-energy-2016-full-report.pdf.
23. Energy Information Administration (EIA) *Mexico*, September 21, 2015, https://www.eia.gov/beta/international/analysis_includes/countries_long/Mexico/mexico.pdf.
24. Dolia Estevez, "Mexico Reverses History and Allows Private Capital into Lucrative Oil Industry," 12/11/2013, http://www.forbes.com/sites/doliaestevez/2013/12/11/mexico-reverses-history-and-allows-private-capital-into-lucrative-oil-industry/.
25. David Agren, "Mexico Opens-Up Petroleum Business to Private and Foreign Companies," Special to *USA TODAY*, December 20, 2013, http://www.usatoday.com/story/news/world/2013/12/20/mexico-petroleum-oil-foreign/4152521/.
26. Ibid.
27. Tracy Wilkinson, "Mexican Political Parties at Impasse over Key Reforms," *Los Angeles Times,* June 19, 2014, http://www.latimes.com/world/mexico-americas/la-fg-mexican-political-impasse-reform-20140619-story.html.
28. David Agren, "Mexico Opens up Petroleum Business to Private and Foreign Companies," Special to *USA TODAY*, December 20, 2013http://www.usatoday.com/story/news/world/2013/12/20/mexico-petroleum-oil-foreign/4152521/.
29. Alexandra Alper and David Alire Garcia, "Mexico's Pemex Signs Cooperation Agreement with Exxon Mobil," Reuters, October 2, 2014, http://www.reuters.com/article/2014/10/02/mexico-pemex-idUSL2N0RX16I20141002?feedType=RSS.
30. Laurence Iliff, "Mexico's Energy Reform Gives Birth to a New Private Oil Company," *The Wall Street Journal*, October 1, 2014, http://online.wsj.com/articles/mexicos-energy-reform-gives-birth-to-new-private-oil-company-1412191958.
31. Gordon Pickering and Robert Gibb, "Mexico Energy Reform: The Other Story," *Pipeline & Gas Journal*, August 2016, Vol. 243, No. 8, https://pgjonline.com/2016/08/01/mexico-energy-reform-the-other-story/.
32. BP, *BP Statistical Review of World Energy June 2016*, http://www.bp.com/content/dam/bp/pdf/energy-economics/statistical-review-2016/bp-statistical-review-of-world-energy-2016-full-report.pdf, last accessed on September 11, 2016.
33. U.S. Energy Information Administration, *China*, EIA, May 14, 2015.
34. IHS Energy, "CNOOC: Government Drivers," September 2014, p. 21.
35. U.S. Energy Information Administration, *China*, EIA, May 14, 2015, p. 4.
36. Ibid.
37. Ibid.
38. BP, *BP Statistical Review of World Energy June 2016*, http://www.bp.com/content/dam/bp/pdf/energy-economics/statistical-review-2016/

bp-statistical-review-of-world-energy-2016-full-report.pdf, last accessed on September 11, 2016.

39. Organization of Petroleum Exporting Countries (OPEC), *Nigeria Facts and Figures*, http://www.opec.org/opec_web/en/about_us/167.htm.

40. U.S. Energy Information Administration, *Nigeria*, EIA, May 16, 2016, p. 3.

41. Ibid., p. 3.

42. Anayo Ezugwu, "Petroleum Industry Bill to be Passed Speedily," Realnews, http://realnewsmagazine.net/politics/petroleum-industry-bill-to-be-passed-speedily-dogara/.

43. Oviemuno Osah, "PIB On Hold Due to Niger Delta Unrest," *BizWatchNigeria*, September 6, 2016, http://bizwatchnigeria.ng/pib-hold-due-ndelta-unrest/.

44. U.S. Energy Information Administration, *Nigeria*, EIA, May 16, 2016, p. 3–4.

45. BP, *BP Statistical Review of World Energy June 2016*, http://www.bp.com/content/dam/bp/pdf/energy-economics/statistical-review-2016/bp-statistical-review-of-world-energy-2016-full-report.pdf, last accessed on September 11, 2016.

46. U.S. Energy Information Administration, *Iraq*, EIA, April 28, 2016, p. 3.

47. Ibid., p. 4.

48. U.S. Energy Information Administration (EIA), "Who are the Major Players Supplying the World Oil Market?" *Energy in Brief*, http://www.eia.gov/energy_in_brief/article/world_oil_market.cfm?src=email, Last updated February 11, 2016.

49. J. William Carpenter, "The World's Biggest State Owned Oil Companies," *Investopedia*, December 17, 2015, http://www.investopedia.com/articles/investing/121715/worlds-biggest-state-owned-oil-companies.asp.

50. David G. Victor, "National Oil Companies and the Future of the Oil Industry," *The Annual Review of Resource Economics*, Vol. 5, (June 2013), p. 445–462.

51. Silvana Tordo, Brandon S. Tracy and Noora Arfaa, *National Oil Companies and Value Creation*, Working Paper 218, Chapter 2, The World Bank (Washington, DC, 2011).

52. David G. Victor, "National Oil Companies and the Future of the Oil Industry," *The Annual Review of Resource Economics*, Vol. 5, (June 2013), p. 445–462.

53. L. E. Grayson, *National Oil Companies*. (Chichester, UK: Wiley, 1981).

54. C. McPherson, *National Oil Companies: Ensuring Benefits and Dealing with Systemic Risks*. In *The Handbook of Global Energy Policy*, ed. A. Goldthau, (Hoboken, NJ: Wiley-Blackwell, 2013), pp. 146–156.

55. David G. Victor, "National Oil Companies and the Future of the Oil Industry," *The Annual Review of Resource Economics*, Vol. 5, (June 2013), pp. 445–462.

56. C. van der Linde, *The State and the International Oil Market: Competition and the Changing Ownership of Crude Oil Assets*. (Boston: Kluwer Academic Publishers, 2000).

57. A. C. Brown, *Oil, God, and Gold: The Story of Aramco and the Saudi Kings*. (Boston/New York: Houghton Mifflin, 1999).

58. P. Stevens, *Saudi Aramco: The Jewel in the Crown*. See Victor et al. 2012b, *Oil and Governance: State-Owned Enterprises and the World Energy Supply*. (Cambridge, UK/New York: Cambridge Univ. Press, 2012), p. 173–233.

59. David G. Victor, D. R. Hults, and M. C, Thurber, eds., *Oil and Governance: State-Owned Enterprises and the World Energy Supply*. (Cambridge, UK/New York: Cambridge Univ. Press, 2012b), p. 3–31.

60. C. Warshaw, *The Political Economy of Expropriation and Privatization in the Oil Sector*. See Victor et al., *Oil and Governance: State-Owned Enterprises and the World Energy Supply.* (Cambridge, UK/New York: Cambridge Univ. Press, 2012b), p. 35–61.

61. David G. Victor, "National Oil Companies and the Future of the Oil Industry," *The Annual Review of Resource Economics*, Vol. 5, June 2013, p. 445–462.

62. Ibid.

63. G. J. Miller, "The Political Evolution of Principal-Agent Models," *Annual Review of Political Science,* Vol. 8, No. 1, 2005, p. 203–225.

64. B. Mommer, *Global Oil and the Nation State.* (UK/New York: Oxford Univ. Press, 2002).

65. David G. Victor, "National Oil Companies and the Future of the Oil Industry," *The Annual Review of Resource Economics*, Vol. 5, June 2013, p. 445–462.

66. N. Victor and I. Sayfer, *Gazprom: The Struggle for Power,* 2012, See Victor et al., *Oil and Governance: State-Owned Enterprises and the World Energy Supply.* (Cambridge, UK/New York: Cambridge Univ. Press, 2012b), p. 655–700.

67. R. Vitalis, *America's Kingdom: Mythmaking on the Saudi Oil Frontier.* (Stanford: Stanford Univ. Press, 2006).

68. Charles Lipson, *Standing Guard – Protecting Foreign Capital in the Nineteenth and Twentieth Centuries.* (Berkeley: University of California Press, 1985), p. 111.

69. Andrew Inkpen and Michael H. Moffett, *The Global Oil and Gas Industry—Management, Strategy & Finance.* (Tulsa, OK: PennWell Corporation, 2010), p. 63.

70. Andrew Inkpen and Michael H. Moffett, *The Global Oil and Gas Industry—Management, Strategy & Finance.* (Tulsa, OK: PennWell Corporation, 2010), p. 57.

71. Ibid., p. 57.

72. Ibid., p. 57.

73. Brian Levy, "World Oil Marketing in Transition," *International Organization* 36. (Winter 1982).

74. David G. Victor, David R. Hults and Mark C. Thurber, *Oil and Governance—State-Owned Enterprises and the World Energy Supply.* (New York: Cambridge University Press, 2012), p. 3.

75. U.S. Energy Information Administration (EIA), "Who are the Major Players Supplying the World Oil Market?" Energy in Brief, http://www.eia.gov/energy_in_brief/article/world_oil_market.cfm?src=email, Last updated on February 11, 2016.

76. *The Wall Street Journal*, "Big Oil's Latest Roadblock: Governments Reduce Access to Aid National Champions amid Scramble for Supply," September 24, 2007, p. C10; Dermot Gately, "What Oil Export Levels Should We Expect from OPEC?" *Energy Journal,* 28 (2007): 151–173; Petter Osmundsen, Klaus Mohn, Bard Misund, and Frank Asche, "Is Oil Supply Choked by Financial Market Pressures?" *Energy Policy,* 35 (2007): 467–474; Dermot Gately, "OPEC's Incentives for Faster Output Growth," *Energy Journal,* 25 (2004): 75–96; Dermot Gately, "How Plausible is the Consensus Projection of Oil below $25 and Persian Gulf Oil Capacity and Output Doubling by 2020?" *Energy Journal,* 22 (2001), p. 1–27.

77. Peter Hartley, Kenneth B. Medlock, III, and Stacy L. Eller, "Empirical Evidence on the Operational Efficiency of National Oil Companies," in *The Changing Role of National Oil Companies in International Energy Markets,* March 2007, http://link.springer.com/article/10.1007/s00181-010-0349-8#page-1.

78. Conglin Xu, Laura Bell, *"OGJ150 Suffered Huge Losses in 2015; Capital Spending, Reserves Down,"* "OGJ100 Posts Lower 2015 Earnings, Higher Oil Production Oil & Gas Journal," September 5, 2016, file:///C:/Users/steve/Desktop/ogjournal20160905-dl.pdf.

79. Oil & Gas Journal, http://www.ogj.com/content/ogj/en/downloadables/survey-downloads/ogj-200/2013.pdf.

80. Conglin Xu, Laura Bell, *"OGJ150 Suffered Huge Losses in 2015; Capital Spending, Reserves Down,"* "OGJ100 Posts Lower 2015 Earnings, Higher Oil Production Oil & Gas Journal," September 5, 2016, file:///C:/Users/steve/Desktop/ogjournal20160905-dl.pdf.

81. Oil & Gas Journal, http://www.ogj.com/content/ogj/en/downloadables/survey-downloads/ogj-200/2013.pdf.

82. Oil & Gas Journal, http://www.ogj.com/content/ogj/en/downloadables/survey-downloads/ogj-100/2013.pdf.

83. Barclays, Global 2013 Capital Spending Outlook, Courtesy of Barclays Research Department.

84. Conglin Xu, Laura Bell, *"OGJ150 suffered huge losses in 2015; capital spending, reserves down,"* "OGJ100 posts lower 2015 earnings, higher oil production Oil & Gas Journal," September 5, 2016, file:///C:/Users/steve/Desktop/ogjournal20160905-dl.pdf.

85. Barclays, E&P Capex: First Double Dip Since '86, January 14, 2016, http://www.oilandgas360.com/ep-capex-first-double-dip-since-86/.

86. Oil & Gas Journal, http://www.ogj.com/content/ogj/en/downloadables/survey-downloads/ogj-200/2013.pdf.

87. Oil & Gas Journal, http://www.ogj.com/content/ogj/en/downloadables/survey-downloads/ogj-100/2013.pdf.

88. BP, *BP Statistical Review of World Energy June 2013*, http://www.bp.com/statisticalreview.

89. Conglin Xu, Laura Bell, *"OGJ150 Suffered Huge Losses in 2015; Capital Spending, Reserves Down,"* "OGJ100 Posts Lower 2015 Earnings, Higher Oil Production Oil & Gas Journal," September 5, 2016, file:///C:/Users/steve/Desktop/ogjournal20160905-dl.pdf.

90. BP, *BP Statistical Review of World Energy June 2016*, http://www.bp.com/content/dam/bp/pdf/energy-economics/statistical-review-2016/bp-statistical-review-of-world-energy-2016-full-report.pdf, last accessed on September 11, 2016.

91. Amy Myers-Jaffe and Ronald Soligo, "The International Oil Companies," The James A. Baker III Institute for Public Policy, November 2007, p. 36–37. http://bakerinstitute.org/files/707/.

92. Ibid., p. 39.

93. Douglass C. North, *Institutions, Institutional Change and Economic Performance.* (Cambridge, UK: Cambridge University Press, 1990), p. 8.

94. Peter Gourevitch, *Politics in Hard Times.* (Cornell University, Ithaca, New York, 1985), p. 54–65.

95. OpenSecrets.org, *Center for Responsive Politics, Influence and Lobbying, https://www.opensecrets.org/lobby/indusclient.php?id=E01, Last accessed on September 17, 2016.*

Chapter 3

Contract Provisions, Bilateral Treaties, Contract Disputes and Expropriations

This chapter examines the most common provisions in oil and gas contracts; the role of domestic and foreign courts and arbitration tribunals; the most common provisions in bilateral treaties; the origins and resolution of contract disputes; the resolution of investment disputes using bilateral treaties; the effectiveness of international courts and tribunals and the causes and frequency of expropriation.

CONTRACT PROVISIONS IN OIL AND GAS CONTRACTS

It is almost inevitable that the parties will disagree about something during the life of a contract. This is even more likely in oil and gas contracts because they may be in operation for ten to twenty years. The disagreement could be whether one of the parties has done what they said they would do, in the way they agreed to do it and within the time frame they agreed to do it. In addition, it is common for oil and gas contracts to include provisions specifying that some issues will be negotiated later. This section summarizes the contract provisions that are most often included in oil and gas contracts to minimize disputes and resolve conflicts.

Good Faith

"Good faith" is a term that is often used with issues pertaining to whether the parties to a contract have done what they said they would do, in the way they promised to do it and within the time frame they agreed to do it. For example:

Excerpt from a Libyan Exploration and Production Sharing Agreements

13.5.2. "shall proceed in good faith to negotiate a gas sales agreement incorporating the principles set forth in Article 13.4."[1]

(This provision is relevant when natural gas is produced from a project that was undertaken with the expectation that it would produce oil.)

Excerpt from an Azerbaijan Agreement

15.2. (cd)–" SOCAR [State Oil Company of Azerbaijan Republic] and Contractor shall diligently negotiate each, such Supplemental Agreement (and the relevant Sales Agreement) in good faith."

Accounting Procedures

In an appendix to the contract there is usually a section titled Accounting Procedures which is referred to in the main contract.

Excerpt from an Appendix on Accounting Procedures

"if any of such methods [of accounting] prove to be unfair or inequitable to the Contractor then the Parties will meet and in good faith endeavor to agree on such changes as are necessary to correct any unfairness or inequity."

Excerpt from an Iraqi Technical Service Agreement

2.3–"Discovered but undeveloped reservoirs, as defined in Annex D, may be developed and produced under this Contract but shall be subject to a separately agreed remuneration fee which the Parties undertake, in good faith, to agree."

Two other phrases frequently used in oil and gas contracts are the obligation to "use reasonable efforts" or to do something or deliver something "in a timely manner." Although domestic law of the host country may provide some guidance on what "reasonable efforts" or "deliver in a timely manner" require of the parties, there will occasionally be disputes whether what has been done constitutes "reasonable efforts" or "timely manner." To reduce the ambiguity of such provisions the contract may include a reference to "generally accepted international standards or practices."[2] However, there can be disputes over the interpretation of terms like "generally accepted international standards or practices" or "good petroleum practice." See the section on environmental provisions below for a definition and example of these terms.

Most disputes arise when one party's interpretation of a provision results in the other party having to spend more money or receive less money than it believes it should under the terms of the contract. An oil or gas contract should always include a section that specifies the procedures the parties will use to resolve disagreements. This is frequently referred to as the "Dispute Resolution" section or the "Arbitration" section and is frequently arranged in a hierarchy.

Excerpt from the Iraqi Model Form Technical Services Contract in Article 37.[3]

> "The Parties shall endeavor to settle amicably any dispute (the "Dispute") arising out of or in connection or in relation to this Contract or any provision or agreement related thereto."
>
> "Where no such settlement is reached within thirty (30) days of the date when one Party notifies the other Party of the Dispute, then the matter may, as appropriate, be referred by the Parties to their senior management for resolution."
>
> "Where no such settlement is reached within thirty (30) days of such referral to management, any Party to the Dispute may refer the matter, as appropriate to an independent expert, or, by giving sixty (60) days-notice to the other Party, refer the matter to arbitration as stipulated hereunder."
>
> "If any Dispute arises between the Parties with respect to technical matters, such Dispute may, at the election of either Party, be referred to an independent expert ("Expert") for determination."
>
> "All Disputes arising out of or in connection with this contract, other than those Disputes that have been finally settled by reference to either senior management or an Expert, shall be finally settled under the Rules of Arbitration of the International Chamber of Commerce by three arbitrators appointed in accordance with said Rules."

In most cases, the parties will first try to resolve disputes themselves, whether the contract requires them to do so or not. Doing so is less expensive and less prejudicial to the relationship, than bringing in outside experts or arbitrators. Only when resolution of the dispute on their own appears impossible, will the parties resort to other dispute resolution mechanisms. The contract may specify the appointment of an expert for technical issues for which there is objective market or accounting data. Disputes that involve more subjective issues of contract interpretation will most likely be referred to a mediator or arbitral tribunal.[4]

Stabilization Contract Provisions

The circumstances in which a contract was negotiated can change during the life of a project in which billions of dollars are invested and which may extend over twenty-five years or more. Government policy, a country's economic development priorities or legal environment can all change after elections or revolutions. Stabilization contract provisions have been used to secure guarantees against changes in the economic policies or legal environment within which a contract was originally negotiated.

Specific changes that a company might seek protection against include: (1) changes in the fiscal regime, for example, increases in existing taxes or royalties, or the imposition of new ones; (2) changes to laws that affect the conduct of petroleum operations, for example labor laws imposing more stringent or additional requirements on worker safety.[5]

Excerpt from a Ghanaian Production Sharing Contract

> 26.2–"...As of the Effective Date of this Agreement and throughout its Term, the State guarantees the Contractor the stability of the terms and conditions of this Agreement as well as the fiscal and contractual framework hereof, specifically including those terms and conditions and that framework that are based upon or subject to the provisions of the laws and regulations of Ghana (and any interpretations thereof) including without limitation the Petroleum Income Tax Law, the Petroleum Law, the GNPC Law and those other laws, regulations and decrees that are applicable hereto. This Agreement and the rights and obligations specified herein may not be modified, amended, altered or supplemented except upon the execution and delivery of a written agreement executed by the Parties. Any legislative or administrative act of the State or any of its agencies or subdivisions which purports to vary any such right or obligation shall, to the extent sought to be applied to this Agreement, constitute a breach of this Agreement by the State."

However, most national governments perceive such stabilization provisions to be an infringement on their national sovereignty.[6] Although stabilization clauses exist in older petroleum contracts, they are becoming increasingly rare. These clauses are often referred to as "freezing clauses" because they attempt to freeze the host country's law in place, at least as it applies to a specific contract.

A more recent alternative to a stabilization provision is the "equilibrium provision." The purpose of an equilibrium provision is like a stabilization provision, that is, it is intended to preserve the overall economic position of the oil company without appearing to infringe the national sovereignty of the host country. If one of the parties is adversely affected by a change in the country's law, both parties agree to pursue changes to the contract that will restore the adversely affected party to the financial position they enjoyed before the law was changed. For example, if the country modifies its tax law and the taxes imposed on the profits of the oil company are increased by 10 percent, the parties would seek to agree on an amendment to some other terms of the petroleum contract to compensate the oil company for the increased tax it must pay.[7] For example:

Excerpts from a typical petroleum contract

"Without prejudice to other rights and obligations of the Parties under the Agreement, in the event that any change in the provisions of any Law, decree or regulation in force in [name of country] occurs subsequent to the signing of this Agreement which adversely affects the obligations, rights and benefits hereunder, then the Parties shall agree on amendments to the Agreement to be submitted to the competent authorities for approval, so as to restore such rights, obligations and forecasted benefits."

"if after the Effective Date, the financial interests of Contractor are adversely and substantially affected by a change to the Law which was in force on the Effective Date, or by revocation, modification or non-renewal of any approvals, consents or exemptions granted to Contractor pursuant to this Contract (other than as a result of Gross Negligence or Willful Misconduct of Contractor or Operator) the Parties shall, within ninety (90) days, agree on necessary adjustments to the relevant provisions of this Contract in order to maintain Contractor's financial interests under this Contract reasonably unchanged.

Environmental Provisions

The Deepwater Horizon oil spill in the Gulf of Mexico in 2010 was a reminder of the environmental risk associated with the exploration and production of oil and natural gas. Although accidents on this scale are rare, all oil and gas projects involve environmental risk. For this reason, most oil and gas contracts contain clauses related to nine major environmental issues: (1) environmental standards; (2) stabilization; (3) environmental impact assessment; (4) access to protected areas; (5) access to water and other natural resources; (6) gas flaring; (7) responding to emergencies and accidents; (8) decommissioning and remediation; (9) liability, indemnity and insurance.[8] Examples of (1), (3), (7) and (9) are presented below:

Environmental Standards

A Cambodian contract defines "good petroleum practice" as:

"Good Petroleum Industry Practices means the standards and practices, and exercise of that degree of skill, prudence and foresight that would reasonably be expected of persons carrying out international petroleum operations, and adherence to generally accepted standards of the international petroleum industry, including sound environmental provisions."[9]

Environmental Impact Assessment

This excerpt is from the Agreement for the Azeri and Chirac Fields in Azerbaijan.

26.4–". . . an environmental baseline study . . . to be carried out by a recognized international environmental consulting firm selected by Contractor, and acceptable to SOCAR [State Oil Company of Azerbaijan Republic]. SOCAR shall nominate representatives to participate in preparation of the study in collaboration with such firm and Contractor representatives."

Responding to Emergencies and Accidents

This excerpt is from a PSA in Ghana.

> "If Contractor does not act promptly so as to control, clean up or repair any pollution or damage, GNPC [Ghana National Petroleum Corporation] may, after giving Contractor reasonable notice in the circumstances, take any actions which are necessary, in accordance with accepted petroleum industry practice and the reasonable costs and expenses of such actions shall be borne by Contractor and shall, subject to Article 17.5 be included as Petroleum Costs,"[10]

Liability and Indemnity

An example of an indemnity clause appears at 19.1 in the LOGIC Standard Offshore Service Contract:

> "19.1 The CONTRACTOR shall be responsible for and shall save, indemnify, defend and hold harmless the COMPANY GROUP from and against all claims, losses, damages, costs (including legal costs) expenses and liabilities in respect of: (a) loss of or damage to property."[11]

In a contract between Slessor and Vecto Gray, clause B13 states that:

> "the parties mutually and irrevocably undertake to release, defend and indemnify each other for damage to any property, and/or injury to/or death of the personnel of the others, arising out of or in connection with the Work, howsoever caused."

Clause 2 in a standard Deeds of Adherence, states that the signatories are obligated to:

> "be solely responsible for and shall defend, indemnify and hold harmless the other Signatories and the other members of their respective Groups against all Claims arising from, out of, or relating to the Services in connection with:
>
> i. personal injury to or sickness . . . ; and
> ii. loss of, recovery of, or damage to any Property . . . ; and
> iii. Consequential loss."[12]

DOMESTIC COURTS, FOREIGN COURTS AND ARBITRATION

When direct negotiations, mediation and the employment of experts do not lead to a resolution of the dispute, the next step is a domestic court proceeding or international arbitration. Unless a contract includes provisions requiring the parties to use an arbitration process, the first step would be to bring the dispute before a domestic court of the relevant country or countries. However, the reality is that in many jurisdictions the court process may not be independent, or may be slow; and international investors generally (not just oil companies) prefer not to take that risk. Although citizens in the host country may find the suggestion that their courts are not impartial or fair offensive, most contracts include a provision specifying arbitration in a court or tribunal outside the host country.

Arbitration and Arbitration Rules

Arbitration is a process for resolving disputes that could not be resolved by some other means. For the parties involved, particularly oil companies, arbitration has two main advantages over a domestic court proceeding. First, the arbitration is not carried out in the country with which the oil company has the dispute, increasing the probability that the outcome will be fair and impartial. Second, arbitration procedures and awards are, at least in theory, confidential and could therefore be an advantage to an oil company because it allows the company to keep the terms and conditions of its original agreements and subsequent settlements private, thereby protecting its proprietary information and its competitive position. The host government may also perceive the confidentiality of arbitration to be an advantage for competitive reasons or when *signing bonuses* or other controversial payments are involved.

Although the arbitration takes place outside the host country, this is a separate issue from the question of which country's law will be used in adjudicating the dispute. For example, if a dispute arises under a petroleum contract in Ghana, then Ghanaian law applies to the contract, and an arbitration process would decide the dispute applying Ghanaian law, even if the arbitration process takes place in the investor's country or a third country.

Typical features of an arbitration provision in a contract include:

- A clause specifying that the arbitration is to be conducted in accordance with the rules of a specific arbitration institution. There are several recognized international arbitration institutions each of which has a set of rules that apply to the arbitration process. The rules most often used are the rules of the ICSID (International Centre for Settlement of Investment Disputes), the rules of UNCITRAL (United Nations Commission on

International Trade Law), and the rules of the ICC (International Chamber of Commerce).
- A clause specifying where the arbitration is to take place. Often, a "neutral" venue is chosen, that is, one that is not in the country with which the contract is made and not in the country in which the oil company or its parent organization is domiciled. Choices of independent venues might include: Paris, London or Stockholm, but as previously discussed the law to be applied (choice of law) is the relevant governing law of the contract, not the law of the venue.

This excerpt is from an Azerbaijan Joint Development and Production Sharing Contract

> ". . . shall be governed and interpreted in accordance with the principles of law common to the law of the Azerbaijan Republic and English law, and to the extent that no common principles exist in relation to any matter, then in accordance with the principles of the common law of Alberta, Canada . . ."

Other provisions include:

- A clause specifying the number of arbitrators that will determine the dispute. Frequently there are three. This allows each party to select one arbitrator and then to jointly appoint a third or for the third to be appointed pursuant to the relevant arbitration rules.
- A clause specifying the language in which the arbitration is to be conducted, usually a major international language that has some historical relationship to the country to which the dispute relates, such as English, French, Spanish or Portuguese.
- A clause specifying who is responsible for paying the cost of the arbitration. Usually, the expense of arbitration is shared equally among the Parties.[13]

This section presented the contract provisions most often included in oil and gas contracts for minimizing and resolving disputes; and explained how these provisions are applied in the resolution of disputes when the arbitration or court proceeding takes place outside the host country. It also demonstrated that there is a well-defined set of procedural rules for resolving disputes when the contracting parties cannot resolve them on their own. The next section reviews the most common provisions contained in BITs and evaluates their effectiveness in preventing and resolving disputes in the international oil and gas industry.

PROVISIONS IN BILATERAL INVESTMENT TREATIES

Contracts are the principal means of documenting the rights and obligations of the parties participating in a specific oil and gas project. Bilateral investment treaties (BITs) are the means of documenting more broadly the rights and obligations of host countries and foreign investors.

Bilateral Investment Treaties

BITs are agreements between governments in which both governments agree to provide certain protections to investments made by nationals of the other country in their country. The most common and important clauses included in bilateral treaties are presented below.

Choice of Law Clauses

In the section on contracts it was noted that oil and gas contracts usually include a clause specifying the use of host country law in the resolution of disputes. BITs also include a provision specifying the use of the law of the host state in international court proceedings and arbitration tribunals, but they also refer to other sources of law. Most bilateral treaties refer to four sources of law: (1) the bilateral investment agreement itself (the treaty); (2) the municipal (domestic law) of the host state; (3) the provisions of the contract relating to the investment between the parties; and (4) general principles of international law. An example of this choice of law clause is contained in the bilateral investment treaty between Argentina and the United Kingdom (1990), which provides at Article 8(4):

> "The arbitral tribunal shall decide the dispute in accordance with the provisions of this Agreement [bilateral treaty], the laws of the Contracting Party [host country government] involved in the dispute, including its rules on conflict of laws, the terms of any specific agreement concluded in relation to such an investment and the applicable principles of international law. The arbitration decision shall be final and binding on both Parties."

BITs usually include an arbitration clause specifying the arbitral body to which disputes may be submitted. Frequently, the parties name the ICSID.

Treaty Titles and Preambles

Most BITs are formally called "agreements for the promotion and protection of investment." In some agreements, the title refers to reciprocal or mutual protection. Treaty titles are usually followed by short preambles, which focus on the objective of promoting and protecting investment, and providing favorable conditions for foreign investment.[14] China-Germany (2003) is an example of a typical preamble:

> "Intending to create favorable conditions for investment by investors of one Contracting Party in the territory of the other Contracting Party,

Recognizing that the encouragement, promotion and protection of such investments will be conducive to stimulating business initiative of the investors and will increase prosperity in both States,

Desiring to intensify the economic cooperation of both States."

National Treatment and Most Favored Nation Treatment

One of the main objectives of international trade and investment law is to limit state actions that discriminate based on the nationality of the foreign individual, entity, good, service or type of investment.[15] In most bilateral investment agreements national and most favored nation treatment (MFN) are combined into one provision. Article 4(2), Chile-Egypt (1999), is an example of this:

> "Each Contracting Party shall accord investments of the investors of [the] other Contracting Party in its territory a treatment which is not less favorable than that accorded to investments made by its own investors or by investors of any third country, whichever is more favorable."

International Standards of Treatment

Standards of treatment based on fairness and equity existed before the development of modern bilateral investment treaties). Fair and Equitable treatment clauses (FETs) used in BITs and other international investment agreements appeared in early international economic agreements, such as the Havana Charter for an International Trade Organization (1948) and the Economic Agreement of Bogota (1948), as well as in the United States Friendship, Commerce, and Navigation treaties. The first use of the FET treatment clause in an international investment agreement appears in Article I of the Draft Convention on Investments Abroad proposed by Hermann Abs and Lord Shawcross in 1959:[16]

> "Each Party shall at all times ensure fair and equitable treatment to the property of the nationals of the other Parties. Such property shall be accorded the most constant protection and security within the territories of the other Parties and the management, use and enjoyment thereof shall not in any way be impaired by unreasonable or discriminatory measures."[17]

The OECD, which was created in 1961, subsequently produced its own draft convention on the protection of foreign property in 1967, titled the Convention on the Protection of Foreign Property, which includes a FET clause like the one in the Draft Convention on Investments Abroad (1959).[18] However, the Convention on the Protection of Foreign Property has never come into force.

The OECD Convention, in its notes and comments to Article 1, makes a clear reference to the source of the standard: "the standard conforms to the 'minimum standard' which forms part of customary international law."[19] Although the 1967 Draft OECD Convention failed to gain sufficient support among OECD countries for adoption as a multilateral convention, its substantive provisions have served as an important model for bilateral investment treaties.[20] By referring to the OECD model and using it systematically, BITs are effectively referring to this standard as defined by the OECD Draft Convention of 1967.[21]

Because most treaties do not define the substantive content of the standard to be applied, but only refer to an unqualified formulation of the standard, or to one qualified by references to (customary) international law, the contemporary meaning of the "FET" standard still depends on interpretations by individual arbitral tribunals, which are not subject to any effective appellate review. This makes the standard vulnerable to inconsistent interpretation, resulting in uncertainty regarding its meaning.[22]

Minimum Standard of Treatment

Unlike the national and MFN treatment standard discussed in the previous section, in which the standard of treatment is contingent on the treatment of some other party or parties, the substantive content of a minimum standard is not determined by reference to the treatment of other investors or investments. Minimum standards of treatment therefore provide a treaty defined baseline or, in the words of one international investment tribunal "a floor below which treatment of foreign investors must not fall, even if a government were not acting in a discriminatory manner."[23] In practice this is still problematic, because there is often disagreement about what constitutes a "minimum standard" of treatment.

Laws Related to Expropriation

International expropriation law is intended to mediate and to the extent possible, reconcile two general principles of international law, (1) that states exercise permanent sovereignty over their territories and natural resources; and (2) that states must respect the acquired rights of foreigners. The exercise of permanent sovereignty means that private property is not inviolable. Unless the state has made specific commitments not to nationalize a specific investor's assets, states have a right to tax, regulate or expropriate an investor's assets provided the state and the expropriation meet four conditions: (1) The expropriation must be for a public purpose, (2) must be done in accordance with due process, (3) requires that it be done in a nondiscriminatory manner and (4) that it be accompanied by prompt and equitable compensation.[24]

The expropriation and compensation provision of Article IV.2.1 of the draft Multilateral Agreement on Investment (MAI) provides:

> "2.1. A Contracting Party shall not expropriate or nationalize directly or indirectly an investment in its territory of an investor of another Contracting Party or take any measure or measures having equivalent effect (hereinafter referred to as "expropriation") except:
>
> a. for a purpose which is in the public interest
> b. on a non-discriminatory basis,
> c. in accordance with due process of law, and
> d. accompanied by payment of prompt, adequate and effective compensation in accordance with Articles 2.2 to 2.5 below"

Three issues in international expropriation law have remained particularly contentious. First, what economic interests can be expropriated? Second, what government measures amount to expropriation? Third, what is the standard of compensation payable upon expropriation? International investment agreements (IIAs) address each of these issues, but are still open to interpretation.[25] Each is examined below.

The range of economic interests that IIAs protect depends on the definition of 'investment,' which most IIAs define broadly. Under customary international law, both tangible property (i.e. land, equipment and inventory) and intangible property (i.e. company shares, dividends, bank accounts, contract rights, intellectual property and goodwill) can be expropriated.

BITs typically contain a provision prohibiting either country from expropriating the investments of nationals of the other country without due process or just compensation or in violation of international law. Most expropriation cases involve a deprivation of a foreign investor's acquired rights and a corresponding acquisition, or appropriation, of those acquired rights by the state or a third party designated by the state, for example, a national oil company.[26] This is usually referred to as direct expropriation.

A deprivation (expropriation) may also occur because a state interferes in the use of the property or the receipt of its benefits by the investor, even though legal title to the property has not been affected. The assumption of control over property by a government does not automatically constitute expropriation requiring compensation under international law. However, a claim of expropriation is warranted whenever events demonstrate that the owner has been deprived of fundamental rights of ownership and it appears that this deprivation is not temporary. This is usually referred to as indirect expropriation and is often described as "equivalent," "tantamount," "*de facto*," "creeping," "constructive," "disguised," "consequential," "regulatory," or "virtual" expropriation.[27] Events in Venezuela, after the election of Hugo Chavez, are an example of creeping expropriation, which is more common than outright expropriation. Case 5 describes the events in Venezuela under Chavez in more detail.

The issue most hotly contested by the claimant and the respondent in expropriation cases is what constitutes 'prompt, adequate and effective compensation' for expropriation under customary international law. The country expropriating the assets would prefer to pay an amount equivalent to the book value of the investment, that is, the difference between the project's total assets and its total liabilities (total assets – total liabilities = book value) based on historical cost. However, the company whose property is being expropriated usually seeks compensation measured by the fair market value of the investment prior to expropriation. The fair market value of the property is usually the greater of the two values and the difference between them is frequently substantial. This is particularly true in the oil and gas industry in which the value of successful wells exceeds their cost; and the fair market value of oil and gas reserves over the long term have tended to increase, until recently.

The decisions of most international courts and tribunals and scholarly writing support the position that under customary international law the country expropriating the assets is required to pay full compensation, measured by the fair market value of the property that has been taken."[28] However, even in those BITs which expressly define equitable compensation as the fair market value, the specific language used in the agreement can have a significant effect on the amount of the compensation the investor receives.

For example, in an arbitration case between the Airport Development Co. and the Republic of Hungary, the bilateral investment treaty stated that "the amount of compensation must correspond to the market value of the expropriated investments at the moment of the expropriation" and that "the amount of this compensation may be estimated according to the laws and regulations of the country where the expropriation is made." A provision written in this manner would be particularly unfavorable to a company whose asset values have been driven down by acts of the host government prior to the announcement of the expropriation, for example, raising tax or royalty rates, denying license renewals or right of transit.[29]

ORIGIN AND RESOLUTION OF CONTRACT DISPUTES

Most disputes are the result of a disagreement over (1) the interpretation of specific terms and conditions in a contract, (2) developments that were not foreseen when the contract was prepared or (3) changes in the host government's policies. Possible reasons for bringing a dispute before an international court or tribunal include (1) the failure of the contract to provide for the specific contingency that has arisen, (2) repudiation of a contract by the host state, (3) a lack of confidence in the fairness of the domestic courts in the host country or (4) the inclusion of a provision in the contract that specifies that an international court or tribunal is to be used if a dispute cannot be resolved by the parties on their own.

Comparison of the Number of Disputes within and outside the Oil and Gas Industry

At the end of 2015, the United Nations Conference on Trade and Development (UNCTAD) listed 772 cases filed with various international courts or tribunals including the ICSID established in 1966, UNCITRAL established in 1964, ICC established in 1919, Stockholm Chamber of Commerce Arbitration (SCC) established in 1917, ICSID/ Additional Facilities (AF) established in 1978, Cairo Centre Rules established in 1979 and the Permanent Court of Arbitration (PCA) established in 1899. Of these 772 cases, 63 were related to the oil and gas industry, or approximately 8.2 percent.[30]

Number of Disputes Compared to the Number of Major Transactions

Between 1996 and 2002, four cases were filed with an international court or tribunal; between 2003 and 2008, thirty-one cases were filed with an international court or tribunal; and between 2009 and 2015, twenty-eight cases were filed with an international court or tribunal. In 2014 and 2015, the total number of new deals (royalty/tax agreements, PSAs, service contracts, joint ventures, acquisitions and divestitures) in the upstream end of the oil and gas industry, were 1,467 and 910, respectively. However, only two cases were filed in an international court or tribunal in 2014 and nine cases in 2015.[31] This suggests that most IOCs and host countries comply with the terms and conditions of the contracts they sign and resolve most of the disputes on their own.

Table 3.1 Comparison of Oil and Gas Disputes and Governance

(*n* = 63)			Ratings made in 2015				
Defendant Country	*Number of Cases (1)*	*Number of Bilateral Treaties in Force (2)*	*Government Effectiveness (3)*	*Regulatory Quality (3)*	*Rule of Law (3)*	*Control of Corruption (3)*	*Average*
Albania	1	35	52	61	41	36	47
Algeria	1	28	34	10	25	32	25
Argentina	8	53	46	13	18	33	28
Azerbaijan	1	34	42	44	31	18	34
Bolivia	2	12	30	20	13	30	23
Burundi	1	5	15	25	17	49	26
Canada	3	30	95	98	95	94	95
Croatia	1	49	74	66	66	62	67
Ecuador	9	16	35	15	13	21	21
Egypt	1	73	20	25	31	32	27
Georgia	2	30	72	79	64	75	73
Grenada	1	2	49	59	50	64	56
India	1	72	45	35	54	39	43
Jordan	1	42	60	55	70	62	61
Kazakhstan	7	36	54	45	34	26	40
South Korea	1	85	87	84	81	70	80
Nigeria	1	15	12	24	12	7	13
Poland	1	60	75	82	77	71	76
Russia	7	59	51	37	26	20	34
Spain	1	73	85	75	80	70	78
Tajikistan	1	22	23	16	15	14	17
Trinidad & Tobago	1	12	64	60	53	34	53
Turkey	1	75	67	66	60	54	62
Uganda	1	7	38	39	40	12	32
Ukraine	3	55	40	29	23	15	27
Venezuela	5	27	10	3	0	5	5
Total	63						

Sources: (1) United Nations Conference on Trade and Development, http://investmentpolicyhub.unctad.org/ISDS/FilterByEconomicSector (2) United Nations Conference on Trade and Development, http://investmentpolicyhub.unctad.org/IIA/IiasByCountry#iiaInnerMenu. (3) World Bank, Worldwide Governance Indicators, http://data.worldbank.org/data-catalog/worldwide-governance-indicators .

Comparison of Disputes and Institutional Effectiveness

Table 3.1 summarizes the sixty-three cases reported by UNCTAD and compares them to the ratings for the quality of governance given to the host country defendants by the World Governance Project.[32] This rating system includes six measures of governance: Voice and Accountability, Political Stability/Absence of Violence, Government Effectiveness, Regulatory Quality, Rule of Law and Control of Corruption. This rating system uses a scale from 0 to 100, in which 0 is the worst possible governance rating and 100 is the best possible rating. These six measures are described below.

Voice and Accountability "captures the extent to which a country's citizens are able to participate in selecting their government, as well as freedom of expression, freedom of association and a free media."

Political Stability and Absence of Violence/Terrorism "measures the likelihood of political instability and/or politically motivated violence, including terrorism."

Government Effectiveness "reflects perceptions of the quality of public services, the quality of the civil service and the degree of its independence from political pressures, the quality of policy formulation and implementation and the credibility of the government's commitment to such policies."

Regulatory Quality "reflects perceptions of the ability of the government to formulate and implement sound policies and regulations that permit and promote private sector development."

Rule of Law "reflects perceptions of the extent to which agents have confidence in and abide by the rules of society, and in particular the quality of contract enforcement, property rights, the police, and the courts as well as the likelihood of crime and violence."

Control of Corruption "reflects perceptions of the extent to which public power is exercised for private gain, including both petty and grand forms of corruption, as well as "capture" of the state by elites and private interests."[33]

Voice and Accountability and Political Stability/Absence of Violence are not included in the analysis below because these two measures are related to individual citizens' rights and personal safety, whereas the other four measures are directly related to the credibility of the government's commitment to policies affecting investors, regulatory fairness and consistency, administration of impartial justice affecting investment disputes and control of corruption influencing investment and investor rights.

Table 3.1 shows that the countries in which the largest number of disputes arose have relatively low ratings on government effectiveness, regulatory quality, rule of law and control of corruption measures (Argentina, Ecuador,

Kazakhstan, Russian and Venezuela). However, there are countries with low ratings on these governance measures which have a small number of disputes (Algeria, Azerbaijan, Bolivia, Burundi, Egypt, Bangladesh, Egypt, India, Nigeria, Kyrgyzstan, Tajikistan, Uganda and Ukraine). This could be partly the result of the smaller number of oil and gas projects in these countries.

Comparison of Administering Institutions

Of the sixty-three cases in the oil and gas sector, forty-two were brought before the ICSID, nine were brought before the PCA, seven were brought before the SCC, two before the London Court of Arbitration and the administering institution in three cases was not known at the time of writing.

Comparison of Arbitration Rules

Of the sixty-three cases, forty-one were arbitrated under the rules of the ICSID, fourteen were arbitrated under the rules of the United Nations Commission on International Trade Law, seven under the rules of the SCC and one under the rules of the ICSID AF.

Length of Time in Arbitration

Of the sixty-three cases brought before an international court or tribunal between 1996 and 2015, one case was in arbitration for less than one year, thirteen were in arbitration for one year, eight were in arbitration for five years, two were in arbitration for six years, three were in arbitration for seven years, five cases were in arbitration for eight years, three were in arbitration for nine years and one was in arbitration for twelve years.

Legal Instruments Referenced in a Dispute

Of the sixty-three cases, forty-four were being arbitrated pursuant to a bilateral investment treaty between the host country and the home country of the investor; thirteen were being arbitrated pursuant to the Energy Charter Treaty; three pursuant to the provisions of the North Atlantic Free Trade Agreement and three pursuant to a bilateral investment treaty and the Energy Charter Treaty. Bilateral treaties are therefore the legal instrument most often referenced in the adjudication of disputes brought before an international court or tribunal.

BILATERAL INVESTMENT TREATIES, LEGAL SYSTEMS AND DISPUTES

In the section on domestic courts, foreign courts and arbitration it was observed that most disagreements between the parties to a contract are resolved by the parties themselves, but in the section on the origins and resolution of contract disputes it was observed that when a disagreement cannot be resolved by the parties, the legal instrument most often referenced in the legal proceedings is the bilateral investment treaty between the country of the claimant and the country of the respondent.

If a country has signed many BITs is it less likely to be involved in disputes in an international court or tribunal? The rationale for this proposition is that countries that have signed many bilateral investment agreements may have a greater respect for the agreements they sign and the rule of law in general.

Number of Bilateral Treaties and the Number of Investment Disputes

In Figure 3.1, Argentina (far right) has fifty-three bilateral investment treaties (BITs) in force and had eight cases brought against it by companies in the oil and gas industry. Russia has fifty-nine BITs in force and had seven cases brought against it. Ecuador has sixteen BITs in force and had nine cases brought against it. Venezuela has twenty-seven bilateral treaties in force and had five cases brought against it. In addition, there are seventeen countries that were involved in one case, yet the number of bilateral treaties they had in force ranged from two to eighty-five.

Figure 3.1 indicates that the number of bilateral treaties a country has signed is not a reliable predictor of how many cases will be brought against it in an international court or tribunal. Possible explanations for this include: (1) one regime may sign many BITs to attract direct foreign investment and a subsequent regime may place more emphasis on wealth redistribution and resource nationalism. In that case, the large number of treaties a country has signed becomes an institutional artifact of the previous regime and has no connection to the policy of the current regime, for example, Venezuela has signed twenty-eight BITs, Argentina fifty-six, Chile fifty, Cuba fifty-nine, Iran sixty-five and Russia seventy-eight, (2) In some countries the decision to sign a large number of BITs may be based more on political considerations than economic considerations, that is, signing treaties expands a country's political network and in the case of larger countries, its "sphere of influence," for example, China has signed 129 bilateral investment treaties, Egypt 100, France 104, Germany 135, India 82, Italy 88, South Korea 90, Netherlands 95, Spain 82, Switzerland 114, Turkey 94, and the United Kingdom 106.

Quality of Country Legal Systems and the Number of Investment Disputes

A more direct way to evaluate the relationship between the quality of governance and the number of investment disputes a country is involved in is to compare the four elements of governance discussed previously and the number of cases a country is involved in. The rational for this proposition is that a country with a fairer and more reliable legal system is less likely to be involved in disputes before an international court or tribunal, because both parties to the contract are likely to have confidence in the domestic court system.

Figure 3.2 compares the number of cases brought against a country in an international court or tribunal and the composite rating (average) a country received on government effectiveness, regulatory quality, rule of law and the control of corruption, where 0 is the worst possible rating and 100 is the best possible rating.

Argentina has a composite rating of twenty-eight and had eight cases brought against it in an international court or tribunal. Ecuador has a composite rating of twenty-one and had nine cases brought against it. Kazakhstan has a rating of forty and had seven cases brought against it. Russia has a rating of thirty-four and had seven cases brought against it. In contrast, Canada has a rating of ninety-five and had three cases brought against it. However, there are a significant number of countries in this sample with relatively low composite ratings and yet only one case brought against them (lower left quadrant). These include Nigeria, Tajikistan, Azerbaijan, Algeria and

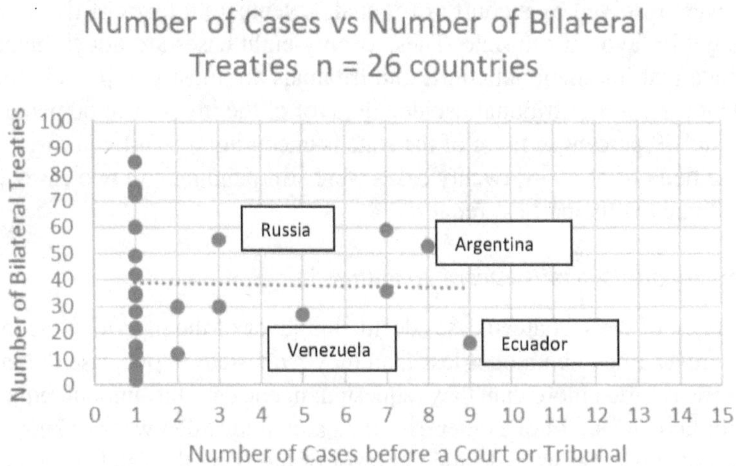

Figure 3.1 Number of Cases vs. Number of Bilateral Treaties. *Source*: Created by the author.

Albania. Another way of looking at this is to note that Canada has a composite rating of ninety-five and had three cases brought against it and Venezuela has a composite rating of just five, and had only four cases brought against it. Figure 3.2 indicates that the composite rating a country received on the quality of its governance and legal system is not a reliable predictor of the number of cases brought against it. The next section examines the effectiveness of international courts and tribunals.

International Courts and Tribunals

The Question of Neutrality

Given the general proposition that institutions develop and are sustained because they reduce risk and expand the range of economic possibilities, it is reasonable to think that international courts and tribunals are at least marginally effective, otherwise there would be little reason for them to exist. The question then is not whether they are effective, but how effective.

As of August 30, 2016, the database of arbitration cases maintained by UNCTAD included 727 cases. Of these 727 cases, fifty-seven are related to the Extraction of Crude Oil and Natural Gas, Six are related to Natural Gas Transport via Pipeline and twenty-two are related to Natural Gas Distribution (local distribution, storage, loading and unloading terminals). The following analysis focuses on the fifty-seven cases and the six cases in the first two categories, that is, the sixty-three cases that we have examined in previous sections of this chapter.

Of the sixty-three oil and gas cases in the UNCTAD database, twenty-eight were resolved by a court or tribunal, seventeen in favor of the investor and eight in favor of the state. These twenty-eight cases are not sufficient to conclude that international courts and tribunals are mostly impartial, but the fact that the court or tribunal decided in favor of the investor in 60 percent of cases and 40 percent in favor of the state is encouraging. Thirteen cases have been settled out of court, twenty cases were still pending and two cases were discontinued at the parties' request.

Claims Amounts versus Award Amounts

In fourteen of the seventeen cases decided in favor of the investors, the investors were awarded an amount less than they requested; in one case the investors were awarded more than they requested; in one case the amount requested was not known; and in one other case the amount awarded was not known. In the eleven cases that were decided in favor of the state, the database indicates that the award, if any, given to the investor is either not available or explicitly reported as zero. Therefore, in twenty-five of the twenty-eight cases, the

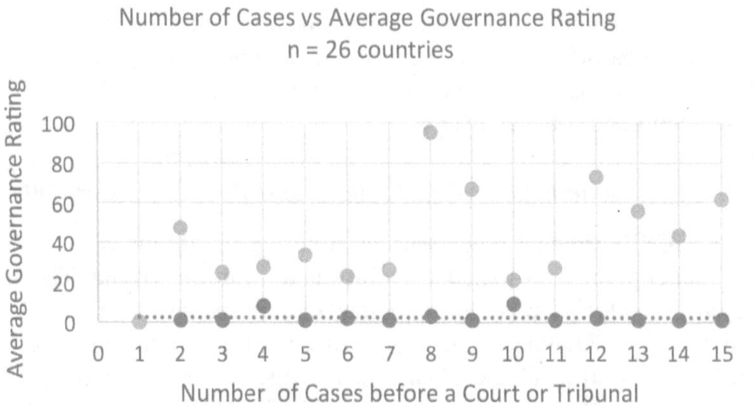

Figure 3.2 Number of Cases versus Average Governance Rating. *Source*: Created by the author.

investors were awarded less than the amount they requested. This is in part, because claimants usually seek compensation equal to the fair market value of the assets that are expropriated or the market value of the breach of contract, but courts are more likely to award an amount equal to the book value, if they decide in favor of the claimant. The origin of the discrepancy between fair market value and book value were discussed in a previous section.

Efficiency of International Courts and Tribunals

One measure of the efficiency of international courts and tribunals is the time between the initiation of the case and the date the court or tribunal rendered its decision. Of the twenty-eight "oil and gas" cases that have been decided by an international court or tribunal, three cases were in arbitration for nine years, two cases for eight years, one case for seven years, two cases for six years, five cases for five years, three cases for four years, eight cases for three years, three cases for two years, and one case for one year. The average time for a decision from an international court or tribunal was 4.6 years.

Of the thirteen cases that were settled out of court or arbitration, two were settled in five years, two were settled in four years, two were settled in three years, three were settled in two years and four were settled in 1 year or less. The average time to settlement was 2.6 years.

Of the twenty cases that were still pending resolution, eight cases have been pending for one year, two cases have been pending for two years, four cases have been pending for three years, one case for five years, one case for seven years, three cases for eight years and one case for twelve years. The average time in adjudication for those cases that are still pending is 3.6 years. A senior official at the ICSID explained this variation by noting that some cases are more complicated than others, some arbitrators are busier than others and some parties challenge every motion and others only challenge basic and critical motions.[34] The next section examines the enforceability of international court and tribunal awards.

Effectiveness of International Courts and Tribunals after the Decision

After an ICSID tribunal renders it decision, the ICSID gives the claimant an official copy of the tribunal's findings and award in writing. The claimant then presents this copy to the defendant and requests compliance with the award. The timeliness and degree of compliance with court or tribunal's awards is difficult to evaluate because the compliance process is less transparent than the arbitration process. UNCTAD does maintain data on "Follow-on Proceedings" related to each case after the court or tribunal's decision has been rendered and the award, if any, has been announced. This could include a request by one of the parties that the decision be annulled or that it be reviewed by a national court. Follow-on proceedings are relatively rare and therefore it might be concluded that most arbitral awards are complied with. There are occasions, however, when an award is renounced, for example Venezuela's rejection of the ICSID's award in favor of ExxonMobil; Venezuela's subsequent withdrawal from the ICSID (see case 5); and Russia's rejection of the award granted by the PCA to the previous shareholders in OAO Yukos. However, a senior official at the ICSID noted that there is an industry consensus that approximately 90 percent of awards are complied with.[35]

CAUSES AND FREQUENCY OF EXPROPRIATION

In the Introduction, it was noted that measuring political risk, that is, the risk of regime change, populist uprisings, civil war and expropriation of assets are hard to predict. For example, Venezuela actively encouraged foreign direct investment in the oil and gas industry prior to 1999, but after Hugo Chavez was elected president of Venezuela the investment climate became increasingly hostile. Vladimir Putin's drive for renationalization of the Russian oil and gas industry did not begin until 2005, even though Putin had been elected prime minister in 1999. The government of Argentina did not expropriate and nationalize Repsol's 57.4 percent interest in YPF until 2012, even though Nestor Kirchner had served as Argentina's president from May 25, 2003 to December 10, 2007, and was thought by many to be a socialist.[36] It was during the presidency of his wife, Cristina Fernandez de Kirchner, who succeeded him, that Repsol's interest in YPF was expropriated.

The Arab Spring that began on December 10, 2010, caught almost all political analysts by surprise as did the chaos that followed in Libya, Tunisia, Yemen, Bahrain and Syria. The sectarian and ethnic divisions that were apparent before the U.S. invasion of Iraq became worse after the U.S. military withdrew from Iraq, and proved to be more disruptive and intractable than anticipated.

Financial and economic crises have been a recurring feature of the global economy. The global financial crises that began in 2007 and continued through 2009, required extraordinary measures by central banks in the developed countries to avert an even worse global recession.

These three types of events create different types of risk. Resource nationalism creates the risk of contract renegotiation and expropriation, but not necessarily the disruption of operations. Military conflicts like those in Syria and Iraq present a risk to the continued operation and potential destruction of energy infrastructure in these countries (i.e. producing wells, pipelines, refineries and export terminals) but not necessarily a risk of expropriation. Excessive financial leverage and insufficient transparency create the risk of financial crises in the capital markets (commercial banks, investment banks, and "non-bank" or "shadow banks"). This section examines the causes and frequency of expropriation from 1960 to 2014.

Table 3.2 demonstrates that the expropriation of oil and gas assets was already underway in the 1960s (20), but increased substantially between 1971 and 1979 (69). Expropriations nearly ceased between 1980 and 2003 and increased again between 2004 and 2011. The average global price of crude oil decreased from $1.90 per barrel in 1960 to $1.80 per barrel in 1970; increased from $2.24 per barrel in 1971 to $31.61 per barrel in 1979; then decreased from $36.83 per barrel in 1980 to $28.83 per barrel in 2003; and increased from $38.27 per barrel in 2004 to $111.26 per barrel in 2011.

Table 3.2 supports Guriev, Kolotilin and Sonin conclusion that the price of oil does not have a significant effect on expropriation risk, but the rate at which the price of oil increases relative to its long-term price trend does affect expropriation risk, that is, the number of expropriations increases when the rate of change in the price of oil increases; and the number of expropriations decreases when the rate of change in the price of oil decreases. However, in the period from 1960 to 1970, there were twenty expropriations even though the average world oil price declined. One possible explanation for this anomaly is that the governments of the oil-producing countries had begun to exert more control over the posted price of oil, but may have been reluctant to raise prices more rapidly because they were uncertain about the consequences of higher oil prices on the world economy and the demand for oil. Therefore, they chose to expropriate the oil and gas fields, which was a more definitive expression of national sovereignty but less risky. (The posted price refers to

Table 3.2 Comparison of the Number of Expropriations and the Price of Oil (Dollars per Barrel and Percent Change)

Year	*Number of Expropriations*	*Annual Average U.S. Crude Oil Price*	*Change in the Price of U.S. Crude Oil in Dollars*	*Change in the Price of U.S. Crude Oil in Percent*	*Annual Average Global Crude Oil Price*	*Change in the Global Price of Crude Oil in Dollars*	*Change in the Global Price of Crude Oil in Percent*
1960–1970	**20**	$2.91–$3.39	$0.48	16.5%	$1.90–$1.80	($0.10)	–5.3%
1971–1979	**69**	$3.60–$25.10	$21.50	597.2%	$2.24–$31.61	$29.37	1311.16%
1980–2003	**3**	$37.42–$27.69	($9.73)	–26.0%	$36.83–$28.83	($7.80)	–21.7%
2004–2011	**12**	$37.66–$87.04	$49.38	131.1%	$38.27–$111.26	$72.99	190.7%
2012	**1**	$86.46	($0.58)	–0.67%	$111.67	$0.41	0.37%
2013	**0**	$91.17	$4.71	5.17%	$108.66	($3.01)	–2.77%
2014	**0**	$93.44	$2.27	2.43%	$104.73	($3.93)	–3.75%
2015	**0**	$48.67	–$44.77	–47.91%	$50.75	–$53.98	–51.54%
2016	**0**	$41.92	–$6.75	–13.87%	$40.19	–$10.56	–20.81%

Sources: 1. Sergei Guriev, Anton Kolotilin and Konstantin Sonin, "Determinants of Nationalization in the Oil Sector: A Theory and Evidence from Panel Data," *Journal of Law, Economics & Organization*, 27 (2) (2011) p. 301. 2. InflationData.com, http://inflationdata.com/Inflation/Inflation_Rate/Historical_Oil_Prices_Table.asp. 3. Y Chats, https://ycharts.com/indicators/average_crude_oil_spot_price. 4. Statista, https://www.statista.com/statistics/206764/forecast-for-west-texas-intermediate-crude-oil-prices/. 5. UNCTAD, http://investmentpolicyhub.unctad.org/ISDS/FilterByEconomicSector.

the price at which a company or country is willing to buy or sell a commodity. In markets where an official exchange does not exist, traders will often refer to the posted price(s) of the major companies or countries trading that commodity. The posted price is therefore like a company's bid and ask price.) The section below presents a brief history of expropriation and privatization in the oil and natural gas industry.

First Attempt at Nationalization—1950 to 1964

In May 1951, after months of unsuccessful negotiations, Iranian premier Mossadegh and his National Front government nationalized the Anglo-Iranian Oil Company (AOIC). The British government responded by blocking Iran's bank accounts in London and the Royal Air Force compelled at least one Panamanian ship to surrender its Iranian oil at Aden.[37] In addition, the seven major IOCs, remained united in their opposition to Iranian nationalization.[38] The boycott of Iran's oil was effective because Iran did not have the technicians needed to operate the nationalized facilities; the refineries needed to refine the crude oil, the tankers to transport it and the distribution system to sell it. Consequently, in 1952 and 1953, Iran sold only 3 percent of the oil it had sold in the years prior to the attempted nationalization.[39] Iran's attempt to nationalize the country's oil and gas industry was ultimately unsuccessful during this period.

The OPEC was founded in 1960, but it did not emerge as a force in setting oil prices until ten years later, during the Tripoli-Tehran crises of 1970 and 1971. Taking control of setting the posted prices in 1960 represented a significant increase in the power of the oil exporting countries and an important structural change in the international oil industry.[40]

The Beginning of the End—1965 to 1971

In 1968, after a successful military coup in Libya, Colonel Qaddafi's Revolutionary Command Council shut down the British and American military bases in Libya. In January 1970, the Revolutionary Command Council demanded a $0.43 per barrel increase in the posted price of oil at a time when the posted price of WTI was $3.35 per barrel. After nine months of negotiation an agreement was reached between the Libyan government and the IOCs. The Libyan government would instead receive a 20 percent increase in royalties and taxes.[41]

In November 1970, the Shah of Iran, surpassed the fifty-fifty profit-sharing terms first established by Venezuela in 1943, by demanding and receiving a tax rate of 55 percent of the profits from the consortium companies in Iran.[42] In 1971, Saddam Hussein nationalized the oil industry in Iraq, increasing its ownership to 100 percent.[43] The period from 1965 to 1971, therefore was a transitional period during which the posted price of crude oil remained essentially unchanged, but the political power of the oil producing countries increased and the power of the IOCs decreased.

The First Oil Crisis—1972 to 1974

At a meeting in Vienna in mid-September 1973, OPEC demanded a new contractual arrangement with the IOCs. The Tehran and Tripoli agreements were terminated just two years after they were signed. On October 5, 1973, Egypt and Syria launched a surprise attack against Israel.[44] On October 17, 1973, the Arab Oil Ministers agreed to an embargo, cutting production by 5 percent from the September level, and agreed to continue cutting production by 5 percent each month until their objective was met, which was, the United States' withdrawal of its support to Israel.[45] Ultimately, the Saudis cut production by 25 percent. OPEC producers had publicly declared solidarity, but some OPEC governments and the Soviet Union increased production. Consequently, the overall reduction in world oil production was just 7 percent,[46] but that was enough to increase the average world oil price from $2.48 per barrel in 1972 to $11.53 per barrel in 1975; and to increase the average U.S. price from $3.60 per barrel in 1972 to $12.21 per barrel in 1975.

The Second Oil Crisis and Nationalization—1975 to 1979

In 1974, Saudi-Aramco was owned 60 percent by the Saudi government, but by 1980 it was owned 100 percent by the Saudi government.[47] In 1976, the nationalization of the Kuwait Oil Company was completed and was owned 100 percent by the government of Kuwait.[48] In 1979, during the Islamic Revolution, the NIOC was nationalized and was owned 100 percent by the government of Iran.[49] The average world oil price increased from $11.53 per barrel in 1975 to $36.83 per barrel in 1980 and the average U.S. price increased from $12.21 per barrel in 1975 to $37.42 per barrel in 1980.

The Era of Privatization—1994 to 2003

The period from1994 to 2003 was dominated by partial or complete privatization of NOCs. For example, between 1989 and 1997, Repsol (a Spanish oil company) was privatized by selling 100 percent of the company's stock to the public. In 1990 and 1991, no privatizations were reported, but in 1992, *Total* (a French oil company) began the process of privatization. The company sold a 30 percent share to the public in 1992 and sold the remaining 70 percent to the public in 1998. In 1993, Oil and Natural Gas Corporation Limited (ONGC, the largest national oil company in India) sold a 16 percent interest to the public. In

1994, OAO Gazprom (a Russian natural gas company) sold a 62 percent interest to the public; and Lukoil (a Russian oil company) became 100 percent privatized. In that same year, a settlement was reached between the government of Peru and a German investor group regarding Peru's expropriation of the undeveloped *Aguaytia* gas field and the producing *Maquia* oil field in central Peru.

In 1995, Indian Oil Corporation Limited (IndianOil) sold 11 percent of its shares to the public. Public ownership of IndianOil was subsequently increased to 21 percent. In 1995, *Enersis S.A.* (a subsidiary of ENI, an Italian oil company) sold a 15 percent interest in the company to the public; and in 2001 public ownership was increased to 70 percent. In 1995, Petróleo Brasileiro S.A. (Petrobras) sold a 49 percent interest in the company to the public. In the period from 1996 through 1999, there were no significant privatizations, but in 2000, two significant partial privatizations occurred. China Petrochemical Corporation (SINOPEC) began the sale of a 24 percent interest in the company to the public; and PetroChina Company Limited (PetroChina) began the sale of a 14 percent interest in the company to the public. In 2001 the Chinese National Offshore Oil Corporation (CNOOC) sold a 29 percent interest in the company to the public and Statoil (a Norwegian national oil company) sold a 38 percent interest to the public.[50] In 2002, no significant privatizations occurred. Therefore, between 1994 and 2003 the oil industry was dominated by privatizations rather than expropriations, even though the average global price of crude oil increased from $15.82 in 1994 to $28.83 in 2003.

Resurgence of Expropriations–2004 to 2011[51]

Beginning in 2004, the frequency of expropriation began to increase again. In 2003, the Russian government presented OAO Yukos, a private sector Russian oil company, with an invoice for $27 billion in back taxes allegedly owed by the company. Yukos was unable to pay the taxes and was subsequently declared bankrupt in a Russian court. Yukos management valued the company at $20 billion. In 2004, its assets were sold to other Russian oil companies and its president was convicted on fraud charges and imprisoned.

In 2006, there were instances of contract renegotiation, contract termination and expropriation: (1) The state of Alaska's agreement with the three oil companies operating on Alaska's North Slope was modified to include an additional 20 percent tax on their profits in Alaska; (2) Occidental Petroleum filed a claim against Ecuador in response to Ecuador's decision to terminate Occidental's exploration and development contract; (3) Bolivia announced that Brazil's state-run company, Petrobras, Royal Dutch Shell PLC, and other private firms would have to transfer ownership of their Bolivian retail gasoline networks to state-owned Yacimientos Petroliferas Fiscales Bolivianos (YPFB) within one month.

In 2007, there were two reports of contract renegotiation and one report of expropriation: (1) Venezuela increased the royalty and tax rates on the oil companies operating in Venezuela. Venezuela's "take" increased from $33.4 billion in 2007 to $38.5 billion in 2008; (2) the Russian government insisted that *Gazprom* be allowed to purchase a 50 percent plus one share interest in Shell's Sakhalin-2 gas field in Russia for $7.45 billion. Shell was forced to give up majority control in one of its most profitable assets after having invested over $6 billion to develop the project; (3) Venezuela expropriated ExxonMobil and ConocoPhillips' assets in Venezuela. ConocoPhillips valued it assets in Venezuela at $4.5 billion and indicated that it would take a charge against the net income for the full amount. The IOCs that were willing to accept Venezuela's terms lost controlling interest in their operations in Venezuela.

In 2010, there was one expropriation and one threat of expropriation: (1) Venezuela nationalized eleven oil rigs owned by Helmerich & Payne (H&P); H&P filed a suit in the Washington D.C. federal district court seeking $32 million in back payments for unpaid services and several hundred million dollars for the value of its eleven drilling rigs, (2) Ecuador threatened to nationalize the oil assets of IOCs that refused to replace their PSAs with fixed-fee contracts.

In 2011, Madagascar Oil of Houston halted operations in Madagascar because it believed its operations might be expropriated. In 2012, YPF, a subsidiary of Repsol, valued at $18 billion, in which Repsol owned a 57.4 percent interest, was expropriated by the government of Argentina.

Absolute and Relative Significance of Expropriation

For the firms whose assets are expropriated, the impact can be substantial and therefore must always be a consideration when making oil and gas investments in a foreign country. One way to measure the relative impact of expropriations on the industry is to compare the value of the properties expropriated in recent years to the total amount of CAPEX in the same years. As already noted, in 2010 there was one expropriation and one threat of expropriation. Venezuela nationalized eleven oil rigs owned by H&P. H&P filed a suit in the Washington, D.C. Federal District Court for $32 million in back payments for unpaid services and several hundred million dollars for the value of its eleven drilling rigs. If the eleven drilling rigs were worth $400 million, the total value of the lawsuit was approximately $432 million. This is a relatively small number compared to the $457.6 billion in CAPEX made by the oil and gas industry in 2010. In 2011, Madagascar Oil of Houston halted operations in Madagascar because it believed its operations might be expropriated. In 2011, the CAPEX made by the oil and gas industry was $556.1 billion. In 2012, Repsol's 57.4 percent interest in YPF, valued at $10.3 billion was expropriated by the government of Argentina. This is a relatively small number compared to the $617.2 billion in CAPEX made by the oil and gas industry in 2012. Expropriation has not been a significant problem since 2011 and is not likely to be a problem in the immediate future given the substantial decrease in the price of oil and gas since June 2014.

NOTES

1. Tim Boykett, Marta Peirano, Simone Boria, Heather Kelley, Elisabeth Schimana, Andreas Dekrout, Rachel O'Reilly, "Oil Contracts – How to Read and Understand a Petroleum Contract," Times Up Press Industriezeile 33b 4020 LinzAustria 2012. http://openoil.net/contracts-booksprint/
2. Ibid.
3. Ibid.
4. Ibid.
5. Ibid.
6. Ibid.
7. Ibid.
8. Kyla Tienhaara, "Foreign Investment Contracts in the Oil & Gas Sector: A Survey of Environmentally Relevant Clauses," *Sustainable Development Law & Policy,* Vol. 11, No. 3 (2011): 15–20, 39–40.
9. Agreement between the Royal Government of Cambodia, Cambodian National Petroleum Agency, Chevron Overseas Petroleum (Cambodia) Limited, Moeco Cambodia Co. Ltd and Woodside South East Asia Pty, Ltd., Article 1.2 Definitions (August 15, 2002).
10. Petroleum Agreement Republic of Ghana; Tullow Ghana Limited; Sabre Oil and Gas Limited; Kosmos Energy. Ghana hc, art. 17.7 (March 10, 2006 (Ghana), http://sec.edgar-online.com/kosmos-energy-ltd/s-1a-securities-registration-statement/2011/03/03/section58.aspx
11. LOGIC, "Standard Contracts for the UK Offshore Oil and Gas Industry: Service Onshore & Offshore," http://www.logic-oil.com/sites/default/files/documents/Services%20Onshore%20and%20Offshore%20Edition%202.pdf
12. LOGIC "Mutual Indemnity and Hold Harmless Deed," (2012), http://www.logic-oil.com/sites/default/files/documents/IMHH%202012%20Deed.pdf, Accessed 12th April 2013.
13. Ibid.
14. Andrew Newcombe and Lluis Paradell, Law and Practice of Investment Treaties: Standards of Treatment. (The Netherlands: Kluwer Law International, 2009), p. 123.
15. Ibid., p. 147.
16. United Nations Conference on Trade and Development, *Fair and Equitable Treatment* – UNCTAD Series on International Investment Agreements II, 2012.

17. Hermann Abs and Lord Hartley Shawcross, "The Proposed Convention to Protect Private Foreign Investment," *Journal of Public Law*, Vol. 9: 115–124, (1960).

18. United Nations Conference on Trade and Development, *Fair and Equitable Treatment*—UNCTAD Series on International Investment Agreements II, 2012.

19. Organization for Economic Cooperation and Development (OECD), *Council Resolution of 12 October 1967 on the Draft Convention on the Protection of Foreign Property, International Legal Materials,* OECD, Vol. 7: 117–143, 1967.

20. Andrew Newcombe and Lluis Paradell, *Law and Practice of Investment Treaties: Standards of Treatment.* (The Netherlands: Kluwer Law International, 2009), p. 30.

21. United Nations Conference on Trade and Development, *Fair and Equitable Treatment—UNCTAD Series on International Investment Agreements II.* (UNCTAD, 2012).

22. Ibid.

23. Andrew Newcombe and Lluis Paradell, *Law and Practice of Investment Treaties: Standards of Treatment.* (The Netherlands: Kluwer Law International, 2009), pp. 233–234.

24. Ibid., p. 321.

25. Ibid., pp. 321–322.

26. Ibid., p. 324.

27. Ibid., p. 327.

28. Ibid., p. 378.

29. Neil Popovic and Alex Lathrop, "Recovery Tactics Outlined for Foreign Takeover Losses," *Oil & Gas Journal,* (25 July, 2007), http://www.ogj.com/articles/print/volume-105/issue-24/general-interest/recovery-tactics-outlined-for-foreign-takeover-losses.html

30. UNCTAD, Investment Policy Hub, Investment Dispute Settlement Navigator, http://investmentpolicyhub.unctad.org/ISDS/FilterByEconomicSector, Last accessed on September 18, 2016.

31. EY, Global Oil and Gas Transactions Review 2015, http://www.ey.com/Publication/vwLUAssets/EY-global-oil-and-gas-transactions-review-2015/$FILE/EY-global-oil-and-gas-transactions-review-2015.pdf.

32. Daniel Kaufmann, Aart Kraay and Massimo Mastruzzi, "The World Governance Project," 2015 Update, http://info.worldbank.org/governance/wgi/index.aspx#home.

33. Ibid.

34. ICSID, Telephone Interview with Senior Official at ICSID on September 23, 2014.

35. Ibid.

36. BBC News, "Analysis: Latin America's New Left Axis," (18 April 2006) http://news.bbc.co.uk/2/hi/americas/4916270.stm.

37. Charles Lipson, *Standing Guard—Protecting Foreign Capital in the Nineteenth and Twentieth Centuries.* (Berkeley: University of California Press, 1985), p. 111.

38. Ibid., 111.

39. Ibid., 112.

40. Theodore Moran, "Managing an Oligopoly of Would-Be Sovereigns: Dynamics of Joint Control and Self-Control in the International Oil Industry," *International Organization*, Vol. 41, No. 4 (Autumn, 1987): 575–607

41. Ibid., p. 580.

42. Ibid., p. 580.

43. Andrew Inkpen and Michael H. Moffett, *The Global Oil and Gas Industry—Management, Strategy & Finance.* (Tulsa, OK: PennWell Corporation, 2010), p. 57.

44. Daniel Yergin, *The Prize—The Epic Quest for Oil, Money and Power.* (New York: Simon & Schuster, 1991), p. 601

45. Ibid, p. 607.

46. Theodore Moran, "Managing an Oligopoly of Would-Be Sovereigns: Dynamics of Joint Control and Self-Control in the International Oil Industry," *International Organization*, Vol. 41, No. 4 (Autumn, 1987): 575–607

47. Andrew Inkpen and Michael H. Moffett, *The Global Oil and Gas Industry—Management, Strategy & Finance.* (Tulsa, OK: PennWell Corporation, 2010), p. 57

48. Ibid, p. 57.

49. Ibid, p. 57.

50. Andrew Inkpen and Michael H. Moffett, *The Global Oil and Gas Industry—Management, Strategy and Finance.* (Tulsa, OK: PennWell Corporation, 2011), p. 62.

51. Oil and Gas Journal website, http://www.ogj.com/_search?q=expropriation

Chapter 4

Financing, Fiscal Regimes, Alliances, Risk Management and Financial Reporting

SOURCES OF FINANCING

In 2013, the global CAPEX in the exploration and production sector of the oil and gas industry was $682 billion and in 2014 it was $688 billion, but decreased to $502 billion (27 percent) in 2015.[1] This decrease in CAPEX was driven by the decrease in the world price of crude oil, for example, the price of WTI decreased from $106 per barrel in June 2014, to $56 per barrel at the end of December 2014, and $29.71 per barrel on February 8, 2016. Global CAPEX was expected to decline further in 2016 to $426 billion.[2] CAPEX is forecast to increase by 5 percent to $448 billion in 2017.[3] Figure 4.1 presents an estimate of the funds (financial capital) raised by oil and gas exploration companies and the source of those funds between 2010 and 2013. Two observations are worth noting. First, the funds raised in 2013 were approximately $850 billion and the CAPEX (not shown) was $682 billion in 2013. First, this implies that the industry raised capital in 2013 in anticipation of even higher capital expenditures in 2014. Second, there is no explicit mention of internally generated cash flow from operations in this chart. Net cash flow from operations would therefore need to be added to these figures to estimate the total sources of financial capital. Nevertheless, Figure 4.1 provides a first approximation of the relative importance of various sources of financing.

Public Equity Investment

Companies can acquire capital and investors can make an equity investment in a company in two ways: publicly or privately. Public companies' shares are traded on a stock exchange and can be purchased by anyone, at any time. The IOCs and many of the major NOCs are traded publicly in New York, London,

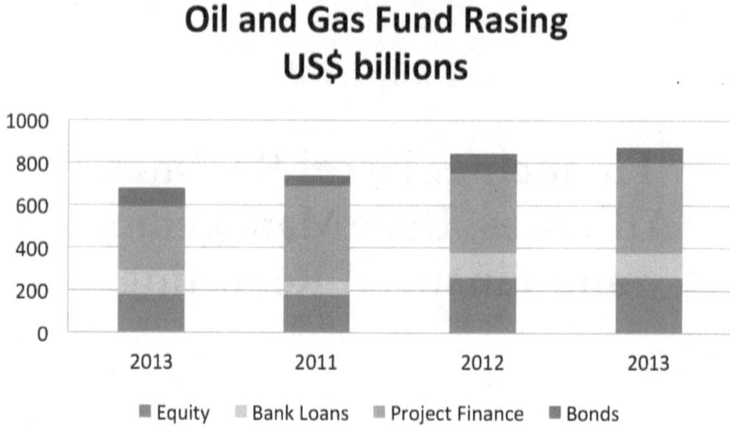

Figure 4.1 Oil and Gas Fund Raising. *Source*: Adapted from E&Y, "Funding Challenges in the Oil and Gas Sector, Innovative Financing Solutions for the Oil and Gas Sector," http://www.ey.com/Publication/vwLUAssets/EY-Funding-challenges-in-the-oil-and-gas-sector/$File/EY-Funding-challenges-in-the-oil-and-gas-sector.pdf

Tokyo or Hong Kong. Many of the medium-sized companies and even the "juniors" in the oil and gas industry are also publicly traded.

Common Stock

Common stock gives its holders ownership in an oil and gas company that is proportional to the number of shares each common stockholder owns. Common stockholders also have a proportional claim on the net income of the firm after the obligations to all the firm's suppliers and creditors have been met. In this sense, the common stockholders bear the ultimate risk of failure or success of the enterprise. In the United States, the net income of the company is taxed at the company level and any dividends that are paid to the stockholders are taxed again at the individual shareholder level.

Master Limited Partnerships

A Master Limited Partnership (MLP) is a limited partnership (or limited liability company) whose units are publicly traded on a stock exchange. If it meets certain qualifications under U.S. tax law, the MLP is a pass-through entity and does not pay income tax at the entity level; income taxes are only paid at the individual partner level. Although not required by law, MLPs generally distribute all their available net income. In addition, each limited partner may record his share of the MLP's depreciation on his tax return.[4] The master limited partners bear the ultimate risk of failure or success of the enterprise.

The first MLP was created by Apache Oil Company in 1981. The number of MLPs in the oil, gas and real estate industries grew rapidly in the 1980s. The U.S. Congress and the IRS became concerned that large numbers of corporations would become MLPs to avoid the corporate income tax. To prevent the widespread adoption of the MLP form, in 1987 Congress passed legislation to define and limit publicly traded partnerships. Congress created Section 7704 of the Internal Revenue Service Code, limiting partnership tax treatment to publicly traded partnerships (PTPs) earning more than 90 percent of their income from a limited number of specific sources.[5]

The popularity of MLPs continued to increase as investors searched for higher yield and energy companies sought to monetize the value of their assets and still maintain control. As of July 2016, there were approximately ninety-six energy MLPs,[6] and the market capitalization for all MLPs was $372 billion.[7] Approximately 70 percent are midstream energy MLPs (pipelines and oil storage), 7 percent are upstream energy MLPs (exploration and production) and the remaining 23 percent are in other industry sectors.[8]

Private Equity Investment

Private equity firms have existed since 1946, but they increased in size and number in the early 1980s. Investors can make an investment in a privately owned company only if the existing owners agree to expand the ownership of the firm. The potential new investors must reach an agreement with the current owners regarding the amount of capital they will contribute and the new investors' proportional ownership in the expanded firm. Private equity capital can come from individual private investors, institutional investors, mutual funds and sovereign wealth funds. Private equity investors frequently hold the investment until it has attained significant value and then sell the company to another investor group or take it public in an initial public offering (IPO).

Hundreds of small oil and gas firms rely on private equity capital to fund their operations. These firms often have significant potential, but have limited access to other forms of capital because they have little operating cash flow and few assets in the beginning, to present to a commercial bank as collateral for a loan. And although advances in technology have improved the odds of finding oil and gas, success is still elusive; three-quarters of all exploration wells are "dry holes," either because there is no oil there or because geologists have been unable to accurately identify the location of the oil or gas.[9] However, 90 percent of the petroleum produced in North America is produced by independent producers, not IOCs. Many of these independent producers are private companies that rely on private equity investment to finance their operations.[10] Private equity groups made significant investments in the oil and gas industry in 2014—104 deals with an aggregate value of

$40 billion. In 2015, the number of private equity deals declined to sixty-four deals with an aggregate value of $37 billion.[11]

Venture Capital

Venture capital financing primarily supports the development of private sector company "startups" that have significant growth potential. The global oil and gas industry includes thousands of startup companies, but their access to venture capital has been limited. In the early 1980s, venture capital investments in the energy and industrial-energy field accounted for more than 20 percent of all venture capital financing, however by 2000, this percentage had declined to 1 percent. Between 2002 and 2008, interest in the energy sector among venture capital firms increased slightly to 3 percent of all venture capital investment in 2007. However, most of this investment was, and still is, focused on clean technologies (biomass, algae and CO_2 capture), not oil and gas exploration and development.[12] For example, investments in clean energy projects increased from $460 million in 2001 to $6.6 billion in 2011.[13]

The oil and gas industry does not attract venture capital for two reasons. First, most new oil and gas technologies (e.g., hydraulic fracturing, horizontal drilling and three-dimensional seismic imaging) require large amounts of capital, often averaging more than $100 million, and ten or more years to reach commercialization and profitability.[14] Second, oil and gas industry technology is perceived as mature and therefore unlikely to produce significant technological innovation.[15]

Debt Financing

Debt financing is available in two forms, commercial bank loans and the issuance of bonds. The most common form of oil and gas financing is senior debt obtained from a bank, or a syndicate of banks (see Figure 4.1), through a revolving credit facility or a term loan credit facility. (A revolving credit facility is a line of credit for which the customer pays an annual commitment fee and is then allowed to borrow money from the lender up to the agreed limit. It is usually used to fund ongoing operations and the amount of the borrowing changes each month depending on the customer's current cash flow needs). If the loan is made to provide an oil and gas producer with working capital or funds to develop existing oil and gas properties, a revolving credit facility is used. However, if the loan is made for the purchase of oil and gas properties, a term loan facility is typically used.

Banks usually secure the loans with a mortgage or deed of trust on the oil and gas properties that are being acquired or developed using the proceeds of the loan. These mortgages or deeds of trust permit the bank to foreclose on the oil and gas properties in the event of a default by the borrower under the credit agreement. If it becomes necessary for the bank to foreclose on the borrower, the bank can sell the oil and gas properties and recover part of the funds it loaned.[16]

Most small and medium-sized firms in the oil and gas industry rely on commercial banks for short-term and medium-term loans to finance their operations, but the availability of long-term commercial bank loans and access to the bond market has historically been limited if these firms specialize exclusively in the exploration and production segment of the industry. In that case, these firms must rely on public and private equity for capital. In contrast, the super-majors (Chevron, ExxonMobil, ConocoPhillips, British Petroleum (BP) and Royal Dutch Shell) and other IOCs have continuous access to commercial bank loans and the bond market because of their vertical integration, more diverse lines of business, the diversity of their exploration and production projects, longer history of operation and their overall size.

Multilateral Financing

Commercial banks (loans) and investment banks (bonds) are the most common sources of debt financing. However, there are other sources of debt financing that are particularly important to developing countries. These multilateral lending organizations include the International Monetary Fund, International Bank for Reconstruction and Development (IBRD), International Development Association (IDA), International Finance Corporation (IFC), Multilateral Investment Guarantee Agency (MIGA), Energy Sector Management Assistance Program, Global Environment Facility and the Carbon Finance Facility.

The World Bank has separated the agencies under its control (Table 4.1) that lend to public enterprises (government and semigovernment organizations) from those that lend to private enterprises (for-profit enterprises). The IBRD and the IDA provide loans to public enterprises, and the IFC lends to private enterprises. There are also regional banks that support investment in the energy industry. These include the Inter-American Development Bank, the European Bank for Reconstruction and Development and the African Development Bank.

Critics have argued that multilateral lending organizations are no longer necessary because of the depth and liquidity of the international financial markets. They also argue that they are highly politicized in their lending policies. Proponents of multilateral lending agencies argue that these institutions provide affordable capital for high-value projects in high-risk countries, projects that would not be financed at affordable interest rates by commercial banks.

Table 4.1 Multilateral Lending Organizations for the Oil and Gas Industry

International Bank Facilities	*Founded*	*Primary Lending and Funding Activities*
International Monetary Fund (IMF)	1944	Provides financial assistance to national governments to help them through serious periods of economic adjustment
International Bank for Reconstruction and Development	1944	Also, referred to as the *World Bank,* it was founded to reconstruct Europe after World War II, and now lends to governments worldwide to support economic and social development
International Development Association	1960	Interest-free loans, credits to finance projects that reduce poverty
International Finance Corporation	1956	Lends directly to private companies without governmental guarantees to promote private enterprise
Multilateral Investment Guarantee Agency	1988	Provides investment guarantees (currency, war, expropriation, breach of contract) to private companies investing in developing countries
Energy Sector Management Assistance Program	1974	Provides advice and analysis but not funds for shaping energy sector development and policy (a joint venture of the United Nations and World Bank)
Global Environment Facility	1991	Provides grants for studies and projects involving national and regional environmental benefits
Carbon Finance Facility	2004	Supports carbon reduction policies and programs in OECD and non-OECD countries which are aligned, using the World Bank Carbon Finance Unit as structure and manager

Regional Development Bank Facilities
Inter-American Development Bank (IDB)
African Development Bank (AIDB)
European Investment Bank (EIB) Nordic Finance Group
OPEC Fund for International Development
Arab Bank for Economic Development in Africa (BADEA)
European Bank for Reconstruction and Development (EBRD)
Asian Development Bank (ADB)
European Union (EU)
Islamic Development Bank
Arab Fund for Economic & Social Development
Source: Hossein Razawi, Financing Energy Projects in Developing Countries, PennWell, 2007. Reproduced in Andrew Inkpen and Michael H. Moffett, The Global Oil and Gas Industry – Management, Strategy and Finance. (Tulsa, OK: PennWell Corporation, 2011

FINANCING STRUCTURES

Centralized Finance System

The centralized finance system is the most common form of financing and ownership structure. It relies on a combination of equity (stock) and debt (commercial bank loans and bonds). In this approach, there is no separation of funds between general operations and the specific projects the company undertakes. When done correctly, it combines a strong central credit rating with a sophisticated intercompany financing system (and the occasional use of project financing) and provides large oil companies with several competitive advantages. These advantages include: low-cost debt and equity, efficient use of the company's cash flow, global tax optimization and facilitation of remittance of its foreign affiliate's cash.

However, financing a project entirely using one company's financial resources is not always feasible or prudent because an individual project may require more capital than one company can afford to risk on a single project, and the political risk of investing in a specific country may be greater than one firm can or should assume. In that case, another financing structure is needed to allocate the risk. These structures include JVs and project finance.

Joint Ventures

A JV is a financing arrangement in which two or more parties agree to pool their financial resources to accomplish a specific objective. In a JV, both participants are responsible for the costs incurred and the profits and losses associated with the venture. The JV is a separate legal entity from the sponsoring companies that are funding the venture. A separate operating agreement is prepared which specifies which company or companies are responsible for actual project development and operation. By jointly funding the enterprise, two or more companies can share the risk associated with a specific project and achieve greater diversification of risk by investing in a larger number of projects. In addition, if a JV includes one or more foreign partners, the number of partners and their individual relationships with the host country may provide some protection against political risk.

Project Finance

IOCs are attracted to large projects, located in difficult environments for two principal reasons. First, large projects provide IOCs the opportunity to use their advanced technology and project management skills as a source of competitive advantage. Second, most NOCs have exclusive access to drilling prospects in their country, particularly onshore. Consequently, the IOCs are frequently forced to operate large projects in difficult environments.

IOCs have often chosen to finance these projects using project financing, because of the large capital requirement, technical complexity and attendant political risk of these projects. In this type of financing structure, the project is primarily financed with commercial bank debt, usually 60 percent or more. Project financing legally separates the project and its funding from the rest of the corporation or corporations that are sponsoring the project. In a project finance arrangement, the lenders rely solely on the assets and cash flow of the project for the repayment of principal and interest. This is significantly different from the centralized corporate finance model, in which lenders rely on the cash flow and financial strength of the entire corporate enterprise for repayment of principal and interest. By employing a project finance structure the participating companies limit their financial and legal risk to the project itself, avoiding exposing any of the sponsoring companies to the financial and legal risk of the project.

Many project finance arrangements in the petroleum industry are structured through a special purpose entity (SPE), an off-balance sheet partnership set up by the company to separate the financial risk of the project from the rest of the corporation. However, it was this type of off-balance sheet financing structure that contributed to the collapse of Enron Corporation. Consequently, internal and external auditors (compliance institutions) are examining these SPEs in more detail and asking for evidence that they serve some business purpose other than to keep debt off the parent company's balance sheet.

Debt levels of 60 percent or higher place a considerable burden on the cash flow from the project for the payment of principal and interest. For the commercial bank or syndicate of banks that provide this capital to be comfortable with the risk they are assuming, the project financing agreement usually contains the following provisions and features.[17]

- The project is established as a new legal entity, separate from the legal and financial responsibilities of its sponsors. This creates a defined environment within which lenders (commercial banks) can evaluate the risks associated with a specific project and it guarantees that the project's cash flow will be used exclusively to repay the principal and interest on the debt of that project. Project financing also protects the sponsor's other assets which are not a part of the project.
- Projects should be long-lived and capital intensive, and they should have a unique purpose. Oil and natural gas pipelines are particularly well suited to this type of financing.
- The project should include cash flow from third-party commitments (customers) that are predictable and reliable. This is usually accomplished

through the establishment of commitments by third parties to "take or pay" for the output from the project. These third-party agreements are usually long-term sales contracts that include price adjustment clauses based on inflation.

- The project should have a finite life, at the end of which all debt and equity will have been repaid.

The use of project financing for high-risk infrastructure construction began with the development of the North Sea oil fields in the 1970s and 1980s. Projects of this type and size were previously financed by issuing utility or government bonds within the framework of a centralized finance system.

Examples of project finance include major liquefied natural gas (LNG) projects such as the Qatar Gas II project (2009), as well as many of the largest individual pipeline investments undertaken in the past three decades, such as the Chad-Cameroon Pipeline (2003) and the Baku-Tbilisi-Ceyhan Pipeline (2005). These projects often combine project financing and equity JVs.[18]

The Chad-Cameroon Pipeline is an example of how project finance can mitigate risk in an oil and gas project. The pipeline is operated by ExxonMobil (40 percent) and sponsored by two other partners forming the consortium, Petronas (35 percent) and Chevron (25 percent). The governments of Chad and Cameroon also have a combined 3 percent stake in the project.[19] The project was begun on October 18, 2000, and completed in June 2003. The financing and construction of the pipeline was made possible by the participation and funding of the World Bank. The Upstream Consortium (ExxonMobil, Petronas and Chevron), provided approximately $2.2 billion (95 percent) of the financing for the pipeline, the World Bank also contributed through debt financing. The IFC provided a loan of about $100 million and helped secure an additional $300 million in private commercial lending. The IBRD provided $92.9 million.[20]

The plans for the project included a revenue management law developed by the World Bank. This law separated the oil revenues given to Chad into four areas, (1) Future Generations Fund, (2) a health, education and other development projects fund, (3) a fund to compensate the Doba region of Chad for the oil extracted from its territory and (4) government reserves. The revenue management law also created the Petroleum Revenue Oversight Committee, which oversees how the oil revenues are spent, and includes members of both the Chadian government and civil society.[21]

The participation of the World Bank and various national export credit agencies provided the IOCs greater assurance that all the parties involved in the project, including the governments of Chad and Cameroon, would comply with their obligations, thereby mitigating political risk.

The Papua New Guinea (PNG) LNG project is another example of a project finance arrangement. The $19 billion PNG LNG project includes gas production and processing facilities that are connected by more than 700 kilometers of onshore and offshore pipeline.[22] The project is led by Exxon Mobil, which is also the operator of the facility. ExxonMobil owns (33.2 percent), Oil Search Limited (29.0 percent), Kumul Petroleum Holding Limited (16.8 percent), Santos Limited (13.5 percent), Nippon Oil and Gas Exploration (4.7 percent) and Mineral Resource Development Company Limited (MRDC 2.8 percent).[23]The company's All Risk Physical Damage insurance provides protection against damage to cargo and construction and losses from terrorism and windstorms.

Structured Project Financing

Structured financing is a more refined approach to project financing. Structured finance involves raising capital and managing risk through the issuance of securities designed to meet the specific risk and reward requirements of a specific investor or group of investors. Structured project finance combines specific securities and derivatives with standard project financing.[24]

Structured project financing reduces the transaction costs created by information asymmetries by limiting the risks a specific group of investors assumes, to those that the investors feel comfortable evaluating. (Information asymmetries exist when investors in a firm's securities do not have the same information or the same level of understanding of the project's risk as the firm's managers, which is usually the case.) Structured project finance also helps the project sponsors and investors ensure that the various types of project risk are allocated to those participants most able and willing to bear them, which increases transaction efficiency and reduces the cost of capital.[25]

Project finance structures can be used to prevent the problems sometimes encountered with customer payments. This is particularly relevant to oil and gas export projects located in countries with unpredictable monetary policies. The bank loan agreements that make these projects possible typically establish *offshore accounts* and *cash waterfalls.* Offshore accounts refer to bank accounts set up in a secure banking center (e.g., New York or London) to which all export customers are instructed to direct their payments. Cash waterfalls refer to written, irrevocable instructions to the bank describing the order in which the cash received is to be paid out to suppliers, short-term creditors, long-term creditors and others. Payments are usually made to suppliers first, then to tax authorities, short-term lenders and long-term lenders. What matters for sponsors, however, is that the project's revenues are kept in a hard currency and in a country with a stable monetary policy, reducing the risk of inconvertibility or devaluation.[26] Figure 4.2 shows a typical project finance structure.

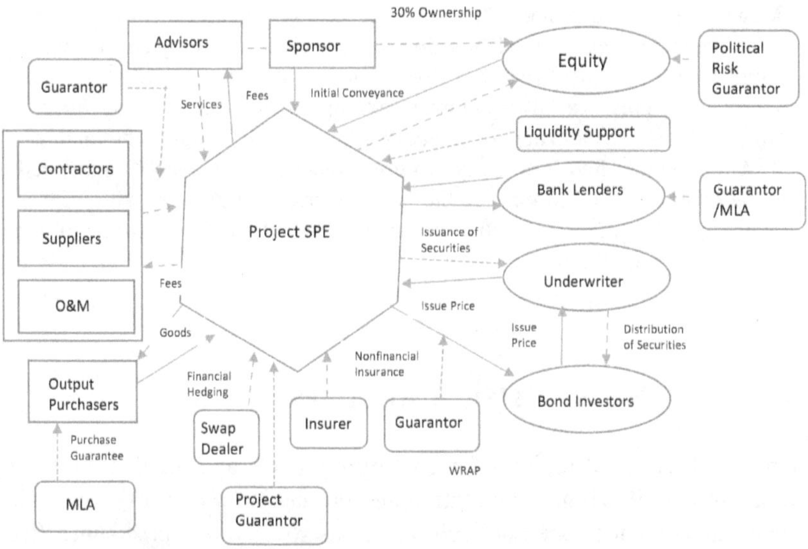

Figure 4.2 Typical Project Finance Structure. *Source*: Adapted from Christopher L. Culp and Paul Forrester, Structured Financing Techniques in Oil and Gas Project Finance, Chapter 21, in Energy and Environmental Finance Law and Taxation: New Investment Techniques, editors Andrea S. Kramer and Peter C. Fusaro (New York: Oxford University Press, 2010)

Project Finance versus Corporate Finance

Project finance has several disadvantages compared to corporate finance. These disadvantages include: (1) the nonrecourse debt used in project financing usually costs more than the corporate debt of the sponsoring companies. (2) Project finance involves third-party costs, such as financial advisers to help structure the financing, engineering reports certifying the quality of the project design, the feasibility of the schedule and the existence of hydrocarbon reserves. (3) Higher legal fees are usually incurred for structuring the deal and preparing the related loan documentation. (4) Structuring a project-financed deal takes longer than a similar deal using corporate funds. (5) The involvement of several third-party lenders and insurance companies reduces the managerial flexibility of the operator of the project and the other companies sponsoring the project. (6) The sponsoring companies may have to disclose proprietary information to lenders.[27]

In a case study prepared by Benjamin C. Esty in 1995, he describes three circumstance in which BP Amoco (now BP) considered employing project financing. These included: (1) mega projects, (2) projects in politically volatile areas and (3) joint ventures with a heterogenous partner.

"Mega project were projects that are large enough to cause 'material' harm to the company's earnings, debt rating, and survival. . . . Projects subject to a high level of political risk, broadly defined as war, strikes, sabotage, loss of property rights, partial or complete expropriation or currency inconvertibility, were also candidates for project finance. . . . In certain joint ventures, BP Amoco might find it necessary to use project finance, even if unjustified on the other criteria... to manage the financial needs of partners with weaker credit capabilities."[28] In all other circumstances, BP Amoco intended to use internal corporate funds.

FISCAL REGIMES

There are three fiscal regimes for controlling assets and the distribution of revenue in the oil and gas exploration and production sector: (1) concession or lease agreement (sometimes referred to as royalty/tax agreement); (2) a PSA or a PSC; and (3) a service contract. When a host country proposes a fiscal regime, it must balance two competing objectives. First, if the host country wants to encourage investment by international oil and gas exploration companies it must provide sufficient incentives for the companies to invest in the country and sufficient opportunity for the companies to recover their costs and earn an appropriate return on investment. Second, the state must balance the interests of the IOCs with the interests of its citizens and export customers.[29]

Concession or Lease (Royalty/Tax Agreements)

In a concession or lease agreement the IOC takes ownership of the hydrocarbons at the wellhead. (The wellhead is the component at the surface of an oil or gas well that provides the structural and pressure-controlling interface for the drilling and production equipment.) The IOC's return on investment is obtained from producing and selling the oil and natural gas; the national governments financial returns are derived from royalty payments and taxes paid by the IOC to the host country. Royalties are usually calculated as a percentage of the project's revenues and is a more stable source of revenue for the host government than revenue based on the project's profit or loss. For example, income taxes or dividend distributions will only be paid if the project or projects are operating at a profit. Although the royalties paid to the host government will also vary with the price of oil, royalties will provide income to the host government every period, whereas as revenue derived from taxing the net income of the IOC on that project will vary more and could be zero if the project is not profitable in a specific period.[30] (See Figure IV.1 in Appendix IV for an example of the revenue split.)

Contractual Systems

The unfavorable terms received by host governments in concession agreements (royalty/tax agreements) in the 1920s and 1930s led host governments and national oil companies to develop so called "contractual systems." The two most commonly used contractual systems are the PSA and the risk service contract. In a contractual system, the state retains ownership of the oil and gas beyond the wellhead. The IOC only takes ownership of oil and gas allocated for recovery of exploration and development costs and allocated from the profit split of volumes for distribution and sale.[31]

Production Sharing Agreements/Production Shàring Contract

Under the terms of a PSA, the IOC is completely responsible for the development of the oil and gas. This includes all aspects of extracting the oil and gas and delivering it to a location for transportation and sale. In most PSAs, the national government receives revenue from three primary sources: (1) royalties, (2) taxes and (3) a share of the oil and gas produced.

When PSAs were first introduced in Indonesia in 1960, the IOCs were not enthusiastic. The prospect of investing large amounts of capital in exploration, equipment and development, but not holding title to the oil and gas, was unsettling for most IOCs. The introduction of PSAs was a major change to the structure of the industry and evidence that the bargaining strength of host governments was increasing and the bargaining strength of IOCs was decreasing. The resistance of the major oil companies to PSAs was overcome when several smaller independent companies without established concession agreements and operations in oil producing countries began signing PSAs to gain access to oil and gas prospects from which they had previously been excluded.[32] However, after the widespread nationalization of IOCs assets in oil and gas producing countries in 1973 and 1974, IOCs embraced PSAs as a risk mitigation technique. The companies forfeited the potential to earn 30 percent returns if oil prices increased, but acquired a greater level of protection against the possibility of expropriation, because the host country's share of revenue increased as the price of oil increased.

In addition, the industry obtained other benefits from the production sharing agreements. These include: the ability to recover invested capital in the first few years of a project, because the IOC's share of production is higher in the first few years of the project; an agreed upon return on investment to the IOC; an increasing share of production when oil prices are declining; and the ability of IOCs to record the oil and gas reserves on their balance sheet. (See Figure IV.2 in Appendix IV for an example of the revenue split in a PSA, also called a PSC.)

Pure Service and Risk Service Contracts

In a *"pure"* service contract or "risk" service contract, the IOC provides some or all the financial capital for exploration and development and is paid by the host government based on a fixed price contract. In a pure service contract the IOC's or service company's revenue is based on the activities it performs, like a fixed price construction contract.[33] The Argentine *Frondizi* contracts of the late 1950s are examples of pure service contracts, named for Argentina's president at the time, Arturo Frondizi. The IOCs were required to drill a specific number of wells per year, per exploration block and in exchange received a fixed dollar amount based on a variety of metrics, including meters drilled, wells completed and ultimately the oil and gas produced per hour.

The terms of a *risk service contract*, provide that if the IOC finds oil or gas, the host country government allows the IOC to recover its costs by selling its share of the oil and natural gas; however, the IOC explores for oil and gas at its own risk and expense.[34] If no oil or gas is found, the IOC bears the cost. The choice of the fiscal regime (concession/lease, PSA, pure service contract or risk service contract) determines the allocation of risk and reward between the parties.

ALLIANCES

Alliances between International Oil Companies and National Oil Companies

The relationship between an IOC and a NOC is usually determined on a project-by-project basis, consequently neither party has any assurance of a broader or longer-term relationship. Some industry analysts have suggested that the interests of IOCs and NOCs would be better served if they formed alliances that extended beyond a single project.[35] They argue that NOCs and IOCs each have unique strengths and these strengths are complimentary. NOCs have the knowledge needed to manage resource and production controls in a manner that the host government perceives to be in the national interest. Furthermore, NOCs are the principal organizations in many developing countries for acquiring new technologies and transferring those technologies to other parts of the domestic economy.[36] Technology transfers can come from a variety of sources, including from other countries' NOCs and from oil field service companies, but technology transfers most often come from the IOCs.[37] IOCs possess substantial technical expertise, project management skills and the ability to define and develop new products and markets.[38]

The arguments for these long-term alliances are not entirely convincing. First, the benefits to be derived from combining the unique capabilities of the NOCs and the IOCs can be achieved whether they work together on a single project or several successive projects. Second, the typical oil or gas project extends over ten to thirty years and therefore is in itself a long-term relationship. Third, NOCs, IOCs and host governments have an incentive to diversify their investment risk among several partners rather than just one or two. Fourth, there are historical and political obstacles to a closer and longer-term relationship between NOCs and IOCs, including the legacy of colonialism, the one-sided nature of the first concession agreements, the potential for resource nationalism and the possibility of sudden political change in the host country. These challenges exist even in a single project and are compounded if the alliance includes several projects with the same partner.

Alliances between National Oil Companies

An alliance between two NOCs may be a more feasible alternative to an alliance between an IOC and a NOC if doing business with an IOC is politically unacceptable in the host country. An alliance between NOCs may also make sense in circumstances where an IOC partner is either difficult to find, or the potential NOC partner is willing to agree to more favorable terms than the IOC. However, in an alliance between two NOCs, some of the advantages that exist in an alliance between an IOC and a NOC are lost, for example, the technical know-how and project management skills of the IOC. In addition, alliances between NOCs may be impeded by current or future political differences between the two countries.

MANAGING COMMERCIAL AND NONCOMMERCIAL RISK

After a project is in commercial operation, it still faces several commercial and noncommercial risks. The commercial risks include: (1) a decrease in revenue as a result of an unexpected decline in product demand or an increase in aggregate product supply that reduces the price of oil and natural gas; (2) an unexpected increase in operating cost; (3) loss of revenue from business interruption or lack of business continuity; (4) property damage; (5) labor disputes and local labor management problems; and (6) liability (workers' compensation, product liability, environmental liability and general liability).

In addition, oil and gas companies operating onshore or offshore of the United States are required to demonstrate that they have sufficient financial resources to pay for damages and the cost of cleanup created by an oil spill.

Noncommercial risks can be divided into two broad categories (financial risk and political risk). The financial risks include changes in interest rates, exchange rates, commodity prices and inflation. Political risk includes: (1) asset impairment (degradation, destruction and expropriation); (2) political violence and war; (3) insecure property rights (contract frustration or abrogation, patent violations, wrongful calling of guarantees, host country failure to honor guarantees and changes in host country laws); and (4) currency inconvertibility and capital flow restrictions.

Managing Commercial Risk

Private Sector Insurers

The risk of a decrease in revenue from an unexpected decrease in product demand, or of an increase in aggregate product supply, and unexpected increases in operating costs are generally not insured against. The oil and gas company's management is responsible for anticipating and adapting to changing economic conditions. Property and casualty insurance is used to provide protection against loss of revenue from business interruption or lack of business continuity; labor disputes and local labor management problems and liability (workers' compensation, product liability and environmental liability).

The purpose and operation of property and casualty insurance is generally understood but the purpose and operation of reinsurance is not as well understood. Reinsurance is the process by which an insurance company transfers a portion of its risk portfolio to other insurers by some form of legal agreement, to reduce the risk that it bears for a specific project. The effect of reinsurance is to spread the risk of completion, operation and maintenance of a project across several insurance companies. In addition to reinsurance for traditional risks like liability, project completion, errors and omissions and business interruption, reinsurers may also provide guaranties and sureties for contractor's performance; provide financing guaranties and sureties for credit risk related to the project (e.g., the risk of nonpayment from contractually committed purchasers of the oil and gas); and provide liquidity support to working capital.

Some of the larger private sector business property and casualty insurers include American International Group (AIG), the Chubb Group of Insurance Companies and Travelers Group. Insurance companies providing property and casualty insurance to exploration and development companies drilling offshore include: Munich Reinsurance Co., Swiss Reinsurance Co. Ltd., Hannover Rueckversicherung AG, Chartis (a subsidiary of AIG), and W. R. Berkley Corporation and Lancashire Group. In response to the Deepwater Horizon accident, major insurance and reinsurance firms have increased the premiums they charge to firms operating drilling rigs in shallow water by 15 percent to 25 percent and for firms operating drilling rigs in deep water by as much as 50 percent.[39]

Mutual Company Insurers

In 1972, sixteen oil companies formed Oil Insurance Limited (OIL). OIL is a mutual insurance company that has fifty members, all of which are oil and gas companies. OIL only insures companies that meet its definition of an energy company. The company's All Risk Physical Damage insurance provides protection against damage to cargo and construction and losses from terrorism and windstorms. Its Control-of-Well-Liability insurance provides protection against perils associated with drilling, for example, well blowouts. Its Third-Party Pollution Liability products insure against liability (including punitive damages) or contractual liability of members for personal/bodily injury, loss of or damage to property arising from a seepage, pollution or contamination incident. The company insures approximately $3 trillion in assets in the oil industry. The company has a per occurrence maximum of $400 million for All Risk Physical Damage. OIL also provides up to $400 million for Control-of-Well-Liability, and $400 million for Third-Party Pollution Liability.[40]

Environmental Liability Insurance

In the early 1960s, a specialty energy insurance market emerged to offer pollution liability coverage for third-party property claims and cleanup and contamination risks, oil well blowouts and redrilling.[41] Insuring the liabilities of vessels was not made compulsory until the adoption of the 1969 International Convention on Civil Liability for Oil Pollution Damage.[42] At the same time, the offshore oil and gas insurance industry began offering insurance coverage for control of well blowouts. Insurers subsequently expanded coverage to include the costs of drilling in deeper water and the cost of redrilling if a blowout occurred.

In 1990, in response to the Exxon Valdez oil spill in Prince William Sound, the United States passed the Oil Pollution and Control Act (OPA). The OPA liability and compensation framework includes a combination of elements that distribute the costs of an oil spill between the responsible party or parties and a trust fund, which is primarily financed through a per-barrel tax on domestic and imported oil in the United States. Responsible parties are liable up to the OPA liability caps of $134 million for offshore oil and gas facilities[43] and $633.9 million for onshore facilities.[44]

The Oil Spill Liability Trust Fund covers costs above the liability limits up to a per-incident cap of $1 billion. However, if the cost of the spill exceeds $1.134 billion for offshore spills and $1.634 for onshore spills, the liability above that amount lies with the party responsible for the accident (usually an IOC or one of its contractors). In the case of the Deepwater Horizon oil spill the responsible party was primarily BP.[45] The final report prepared by the National Commission on the BP Deepwater Horizon Oil Spill and Offshore Drilling mentioned a September 2010 announcement from the offshore oil insurance company, Munich Re advertising environmental coverage in the $10 billion to $20 billion range.[46] However, this is the exception rather than the rule and there is no record of insurers underwriting policies of that size or oil companies being willing to pay the insurance premiums for coverage of that magnitude. Consequently, most exploration and production companies are self-insured for environmental liability, for example, BP in the Deepwater Horizon accident.

Managing Financial Risk

Financial risks include changes in interest rates, currency exchange rates, commodity prices and inflation rates. Financial risks can be managed through a combination of derivatives and long-term purchase agreements. Derivatives include commodity futures, forward contracts, options and swaps. Long-term purchase agreements are usually negotiated prior to the development of an oil or gas field to secure a market for the future output from the field.

Managing Political Risk

Political risk insurance (PRI) provides one means of recovery for companies whose foreign investment has been expropriated or whose financial interests have been damaged by a host government or its citizens.[47] The first PRI policies were issued by the United States after World War II, to encourage private investment in Western Europe. Private sector insurers also offer PRI coverage in developing and developed countries and for various durations. Some of the larger private PRI issuers are Lloyd's of London, Zurich Financial Services Group, Sovereign Risk Insurance Ltd., AIG, and Chubb Corp. In addition, the United States Overseas Private Investment Corporation (OPIC) and World Banks's MIGA provide PRI. However, most government sector providers of PRI are national export credit agencies, which insure short-term export credit/trade transactions. Examples include the U.S. Export/Import Bank, Export Development Canada (Canada), and Eksport Kredit Fonde (Denmark).

Overseas Private Investment Corporation

In 1971, OPIC, a wholly owned U.S. government corporation was established to provide direct financing and PRI for projects in developing countries.[48] On September 30, 2013, OPIC's total worldwide exposure was $18 billion of which 71 percent was related to financing (i.e., loans), 12 percent to PRI and 17 percent to investment funds (equity investments).[49] On September 30, 2014, OPIC's total exposure was $18.02 billion, of which $3.05 billion was risk under OPIC's PRI Program and $14.97 billion under OPIC's Credit Reform program.[50] As of September 30, 2015, OPIC had combined total insurance in force of $19.93 billion.[51] The "loan limit" is $250 million per project and the "house limit" for lending and insurance combined is $400 million. OPIC provides protection against (1) war, civil strife, coups and other acts of politically motivated violence; (2) terrorism; (3) expropriation; and (4) abrogation, repudiation and/or impairment of a contract or other improper host government interference. Companies can purchase protection against whatever risks they choose, that is, the insurance menu is "a la carte"; in addition, a company can purchase protection against "attenuated losses" or "trade-related losses," that is, losses created by a supply disruption between facilities on which its operations depend.

Dispute Resolution Procedures and Awards

The insurer usually waits until an international court or tribunal has rendered its decision and award, before it considers the plaintiff's claim for compensation under a PRI policy. However, while the case is still in arbitration, the insurer will help the parties in the dispute to reach an out-of-court settlement. If the defendant pays the claimant after either of the parties has agreed to a settlement or the court renders its decision and award, then the matter is ended. If the defendant does not pay the claimant, then the insurance company pays the claimant and the insurance company then seeks compensation from the defendant. A senior official at OPIC estimated that in the latter case OPIC recovers 95 percent of what it is owed by defendants. This official estimated that the recovery rate in the private sector of the PRI industry is about 35 percent. He explained the difference in recovery rates by noting that: (1) OPIC can use the full force of the U.S. government departments and agencies (Department of State, Department of Commerce, Department of Justice) to enforce its claims for compensation from the host government; (2) OPIC lends to and insures projects in countries where no lender or insurer in the private sector is willing to lend, so a defendant country is not likely to want to offend either OPIC or the U.S. government by refusing to pay what it owes. If the country refuses to compensate OPIC, OPIC will most likely exit the country and not finance or insure any other projects in that country.[52]

OPIC plays a dominant role in the PRI market and its standard policy is available on the internet. OPIC policies provide coverage for "total expropriation" and cover an act or series of acts by the host government that violate international law or materially breach local law and directly deprive the insured of fundamental rights in the insured investment. However, the standard OPIC policy does not further define when an expropriation constitutes a violation of international law therefore OPIC applies the general principles of international law previously discussed.[53] That is, an expropriation violates international law when it is not for a public purpose, is discriminatory, is not accomplished by due process and is not followed by just compensation.

A senior official at OPIC noted that regulatory risk and other types of "partial takings" are becoming more common than outright expropriation and that this type of event is harder to place a dollar value on.[54]

These "partial takings" refer to the creeping or disguised expropriation discussed earlier and examined in more detail in case 5. He also noted that OPIC considers the existence of a bilateral investment treaty between the host country government and the United States very important to OPIC's decision to insure a project. OPIC also places importance on a country's signature and ratification of the New York Arbitration Convention (established in 1958) and a country's membership in the ICSID.

The Multilateral Investment Guarantee Agency

An official at MIGA suggested that the amount of PRI in the oil and gas sector is small for several reasons (1) IOCs have their own captive insurance companies and therefore can self-insure; (2) Oil companies place a high value and trust in the quality of the relationship they have with the host government's leaders; (3) infrastructure projects (electrical power generation plants, railroads and dams) usually require more capital than all but the largest oil and gas projects; (4) Developing countries more often need help insuring infrastructure projects like electrical generation plants than oil and gas projects; and (5) foreign investors usually own only a small part of a given exploration block, thereby limiting their risk.[55]

This official also observed that although the existence of a bilateral treaty between countries is a consideration in deciding to insure a project, MIGA does not consider it critical, because MIGA will only insure projects in which the contract between the parties includes a provision for mediation and arbitration in the event of a dispute. In addition, MIGA only insures projects in countries that are members of MIGA and all members of MIGA sign a "Legal Protection Agreement" which guarantees MIGA the same rights as the country's "most favored" partners.

This official also noted that the basis of compensation varies depending on the nature of the loss. In the case of (1) expropriation the compensation is based on net book value; (2) in the case of breach of contract MIGA requires mediation by a panel of three experts, and if that fails then arbitration by an international court or tribunal; (3) in the case of inconvertibility or transfer restrictions the amount of money is known; (4) in the case of property damage, the compensation is based on the cost of restoring the property to its prior condition. MIGA deposits the money in an offshore account for the claimant and the claimant receives compensation in a currency that is convertible to another currency.

Use of Political Risk Insurance in the Oil and Gas Industry

On August 2015, MIGA had $12.5 billion of PRI in force and approximately 9 percent of that was related to PRI in the oil and gas industry. [56] On September 2015, the total amount of PRI in force related to foreign direct investment was $258.1 billion.[57] If other political risk insurers (OPIC and private sector insurers) insure in roughly the same proportion, then the total PRI in force in the oil and gas sector at the end of 2015 was approximately $23.2 billion ($258.1 billion x 0.09 = $23.2 billion). However, in August 2014, the estimated total value of all global oil and gas assets was $4.65 trillion.[58] The approximate oil and gas assets insured as a percentage of global oil and gas assets is therefore, 0.5 percent ($23.2/$4,650 = 0.5 percent). These figures indicate that the use of PRI by the oil and gas industry is small compared to the oil and gas assets in place. Although the average global price of oil declined from $104.73 per barrel in 2014 to $40.19 per barrel in 2016, the percentage of oil and gas assets that are insured is still small.

Coverage Limits

OPIC and MIGA offer maximum coverage limits of $250 million and $220 million, respectively. The limits offered by private insurers range from $85 million at AIG to $125 million at Sovereign Risk Insurance Limited. However, political risk insurers often form consortiums to increase the total limits available to the policyholder and to diversify their own risk.[59]

FINANCIAL REPORTING AND OPERATIONAL TRANSPARENCY

Financial reporting quality and operational transparency refer to the extent to which the published financial statements and operational data of a company provide accurate and complete information about the firm's performance. Financial reporting includes financial statements showing, (1) revenue, expenses, net income, cash flow, assets and liabilities; (2) the financial terms and conditions of specific transactions (purchase price, sales price and due dates); (3) legal disputes and potential liability. Operational transparency includes disclosure of the countries in which a company operates, its choice of partners, environmental compliance, wells drilled, new discoveries, reserve depletion rates, and total oil and gas reserves. In theory, high-quality financial reporting and operational transparency should increase the efficiency of markets by reducing transaction costs, monitoring costs and surrounding uncertainty; and the costs of asymmetric information.

Several empirical studies have concluded that higher-quality financial reporting does increase the efficiency of capital markets.[60] The efficiency of capital markets can be defined in two ways. It can be defined in terms of transaction costs, monitoring costs and surrounding uncertainty, that is, the lower these costs are for a given transaction the more efficient the market is said to be. Market efficiency can also be defined in terms of the cost and consequences of asymmetric information, bounded rationality and the adverse effects of opportunism. Understood in this sense, the lower the cost of capital to borrowers for a given level of investment risk, the more efficient the market is said to be.

The studies referred to above, all evaluated the quality of financial reporting and market efficiency in the context of buying and selling financial securities on an organized exchange. They did not relate the quality of financial reporting and market efficiency to foreign direct investment; and it cannot be assumed that what is true in organized financial markets is true of foreign direct investment. However, the largest part of the investment in the oil and gas industry is made in the form of either domestic or foreign direct investment. The question therefore remains whether the quality of financial reporting in a country affects the amount of foreign direct investment inflows to that country in general and to the oil and gas sector.

ESTIMATING AND VALUING RESERVES

Estimating and valuing reserves is the most difficult reporting challenge because future production can only be approximated given the technical, geologic and pricing uncertainty surrounding oil and natural gas. The Society of Petroleum Engineers and World Petroleum Council jointly developed a reserve assessment methodology to address the technical uncertainty of estimating oil and gas reserves. This methodology uses a system that classifies oil and gas reserves into three categories: (1) prospective (undiscovered reserves), (2) contingent (subcommercial reserves) and (3) commercial reserves. This system includes both conventional and unconventional deposits (shale oil, shale gas and tar sands). When oil prices rise and extraction technology improves, conventional and unconventional oil reserves are reclassified from subcommercial to commercial reserves.[61] In 2015, the substantial decrease in oil prices led to some reserves previously classified as commercial reserves being reclassified as contingent reserves.

Reporting Oil and Gas Reserves

The first tier of reporting is performed by individual oil and gas companies (NOCs, IOCs and domestic oil companies). The second tier of reporting is carried out by public information agencies (the IEA, the U.S. Energy Information Administration (EIA), and the OPEC Secretariat and private companies [BP and the Oil & Gas Journal]). Owen et al., concluded that in some cases, these second-tier organizations acknowledge sources of reporting error, but in general they reproduce the data obtained from the first tier reporting organizations with only small adjustments to account for differences in oil grades.[62] They suggest that second-tier sources make more optimistic estimates than independent analysts because the second-tier reporting organizations do not question the estimates made by the first tier, possibly because they consider such questions politically sensitive and diplomatically offensive.[63] However, a senior official at the Department of Energy believes that the IEA and the EIA make a diligent effort to evaluate first-tier estimates and where necessary, reduce them.[64]

At the end of 2013, BP and the Oil & Gas Journal estimated world proved oil reserves (*conventional and unconventional*) at 1,687.9 billion barrels[65] and 1,644.5 billion barrels[66], respectively. At the end of 2014 and 2015, BP estimated world proved oil reserves (*conventional and unconventional*) at 1,700.0 billion barrels and 1,697.6, respectively. In October 2016, the EIA estimated proved reserves at 1,660.8 billion barrels.[67] BP, the Oil & Gas Journal, and EIA are all second-tier reporting organizations whose estimates may be optimistic.

Practical Relevance of Reserve Estimates

In the short term, current oil inventories, that is oil above ground, is the most important determinant of spot (current) oil prices and the level of CAPEX expended in the search for new oil and gas reserves. In the long-term, however, estimates of proved and probable reserves below ground play the most important role in determining the price of oil and CAPEX. If the actual proved and probable reserves turn out to be less than those estimated, the current price of oil may be lower than is justified by long-term supply and demand. The problem of underinvestment could be made worse by the fact that it takes more time and more evidence to convince oil and gas company executives that an oil price increase will persist than it does to convince them that an oil price decrease will persist. That is, an increase in oil prices does not stimulate exploration expenditures in the short run, but a decrease in oil prices will cause an almost immediate reduction in exploration expenditures.[68] This could result in underinvestment in exploration and development and could set the stage for another rapid increase in the price of oil when it becomes evident that reserves are overstated and capital spending has been too low to meet oil demand.

NOTES

1. Barclays, "E&P Capex: First Double Dip Since '86," January 14, 2016, http://www.oilandgas360.com/ep-capex-first-double-dip-since-86/
2. Ibid.
3. Oil & Gas Journal, "Barclays: Global E&P spending to trend up 5 percent in 2017," September 7, 2016, http://www.ogj.com/articles/2016/09/barclays-global-e-p-spending-to-trend-up-5-in-2017.html
4. Internal Revenue Service Code, I.R.C. § 7704 (2006).
5. National Association of Nationally Traded Partnerships, *National Limited Partnerships 101: Understanding MLPs*, October 4, 2013, http://www.naptp.org/documentlinks/Investor_Relations/MLP_101.pdf
6. Dividend Yield Hunter, "Master Limited Partnerships," http://www.dividendyieldhunter.com/master-limited-partnerships-alphabetical/#top
7. Yorkville Capital Management, http://www.yorkvillecapital.com/asset-class-overview.aspx
8. Dimitra Defotis, "The New MLP Landscape," Barron's, Feb. 23, 2013, http://online.barrons.com/news/articles/SB50001424052748704103204578314250446774898
9. Jad Mouawad, "Wildcatters Find their Niche in the Oil Industry," *The New York Times*, May 19, 2007.
10. Oil & Gas Journal, "Private Equity Investments Crucial to Energy Industry," O&GJ, December 1, 2009.
11. Ernst and Young, "Capitalizing on Opportunities – Private Equity Investment in Oil and Gas," http://www.ey.com/Publication/vwLUAssets/ey-capitalizing-on-opportunities/$FILE/ey-capitalizing-on-opportunities.pdf
12. Andrew Inkpen and Michael H. Moffett, *The Global Oil and Gas Industry – Management, Strategy and Finance.* (Tulsa, OK: PennWell Corporation, 2011), p. 62.
13. Raphael Rosen, "Why Venture Capital is Weak Fuel for Clean Energy Startups," Forbes, http://www.forbes.com/sites/energysource/2013/02/13/why-venture-capital-is-weak-fuel-for-clean-energy-startups/#472eb44050b9

14. Robert W. Shaw, Jr. "Energy Venture Capital: The New Wave," Arete Corporation, in Energy Venture Capital Best Practices. (Boston: Aspatore Books, Inside the Minds series, May 2006), p. 8.

15. Rolf Wustenhagen and Tarja Teppo, *What Makes a Good Industry for Venture Capitalists? Risk, Return, and Time as Factors Determining the Emergence of the European Energy VC Market.* (University of St. Gallen, IWO Discussion Paper No. 114, October 2004).

16. Jeffrey S. Munoz, "Financing of Oil and Gas Transactions," *Texas Journal of Oil, Gas and Energy Law*, Vol. 4, No. 2 (2009), p. 223–267.

17. Andrew Inkpen and Michael H. Moffett, *The Global Oil and Gas Industry—Management, Strategy and Finance.* (Tulsa, OK: PennWell Corporation, 2011).

18. Ibid.

19. Martin, J. Paul, "*Chad Cameroon Oil Pipeline Project, a Study Tool and Case Study,*" *Columbia University.*

20. Chad-Cameroon Oil Pipeline: A Case Study from New Era, Center for Energy Economics, University of Texas at Austin

21. Catholic Relief Services, "*Catholic Relief Services Statement on the Chad-Cameroon Oil Pipeline,*" http://web.archive.org/web/20140318155730/http://pwyp-dev.gn.apc.org/sites/pwypdev.gn.apc.org/files/crs_chad_091003.pdf

22. ExxonMobil, PNG LNG, https://pnglng.com/About/Project-overview

23. ExxonMobil, PNG LNG, https://pnglng.com/About/Co-venturers

24. Andrew Inkpen and Michael H. Moffett, *The Global Oil and Gas Industry—Management, Strategy and Finance.* (Tulsa, OK: PennWell Corporation, 2011).

25. Ibid.

26. Stephen Arbogast and Praveen Kumar, *Financing Large Energy Projects*, Chapter 13, in Betty J. Simkins and Russell E. Simkins, *Energy Finance and Economics—Analysis and Valuation, Risk Management, and the Future of Energy.* (Hoboken, NJ: John Wiley & Sons, Inc., 2013).

27. Benjamin C. Esty, *Modern Project Finance—A Casebook.* Chapter 7 (John Wiley & Sons, 2004), p. 144–145.

28. Ibid., p. 148–149.

29. Andrew Inkpen and Michael H. Moffett, *The Global Oil and Gas Industry—Management, Strategy and Finance.* Chapter 6 (Tulsa, OK: PennWell Corporation, 2011), p. 217.

30. Ibid., p. 222.

31. Ibid., p. 223.

32. Ibid., p. 225.

33. Ibid., p. 226.

34. Ibid., p. 226.

35. Robert A. James, *Strategic Alliances between National and International Oil Companies*, Working paper #104, (Freeman Spogli Institute for International Studies, October, 2011), p. 9.

36. Ibid., p. 12.

37. Peter A. Nolan & Mark C. Thurber, *On the State's Choice of Oil Company: Risk Management and the Frontier of the Petroleum Industry.* (Stanford University

PESD Working Paper No. 99, 2010), available at http://pesd.stanford.edu/publications/on_the_states_choice_of_oil_company_risk_management_and_the_frontier_of_the_petroleum_industry/

38. Robert A. James, *Strategic Alliances between National and International Oil Companies.* (Working paper #104, Freeman Spogli Institute for International Studies, October, 2011), p. 12.

39. Moody's Investor Service. "Deepwater Horizon Losses Hit Insurers and Reinsurers," Credit Global Research, June 30, 2010. https://www.moodys.com/research/Moodys-Deepwater-Horizon-Losses-Hit-Insurers-and-Reinsurers--PR_200382

40. Oil Insurance Limited, https://www.oil.bm/

41. Rawle O. King, *Liability and Financial Responsibility Issues Related to Offshore Oil Production*, Senate Committee on Energy and Natural Resources, May 25, 2010, http://www.energy.senate.gov/public/index.cfm/files/serve?File_id=cfe7ede3-9f52-03a2-6432-85f94c70d801

42. International Convention on Civil Liability for Pollution Damage*, 1969*, located at: [http://www.imo.org/conventions/contents.asp?doc_id=660&topic_id=256].

43. Bureau of Ocean Energy Management, "BOEM Adjusts Limit of Liability for Oil Spills from Offshore Facilities," BOEM, Press Release, http://www.boem.gov/press12112014/

44. The United States Coast Guard, The National Pollution Funds Center, "Limits of Liability," https://www.uscg.mil/npfc/Response/RPs/limits_of_liability.asp

45. Jonathan L. Ramseur, Liability and Compensation Issues Raised by the 2010 Gulf Oil SpillMarch 11, 2011, http://assets.opencrs.com/rpts/R41679_20110311.pdf

46. Munich Re, Press Release, September 12, 2010, http://www.munichre.com/en/media_relations/press_releases/2010/2010_09_12_press_release.aspx

47. Neil Popovic and Alex Lathrop, "Recovery Tactics Outlined for Foreign Takeover Losses," *Oil and Gas Journal,* (25 June, 2007), http://www.ogj.com/articles/print/volume-105/issue-24/general-interest/recovery-tactics-outlined-for-foreign-takeover-losses.html

48. Overseas Private Investment Corporation, *OPIC Annual Report 2013*, p. 2, http://www.opic.gov/sites/default/files/files/OPIC_AR2013_final.pdf

49. Ibid.

50. Overseas Private Investment Corporation, *OPIC Annual Report 2014*, p. 9, https://www.opic.gov/sites/default/files/files/opic-financials-0302.pdf

51. Overseas Private Investment Corporation, *OPIC Annual Report 2015*, p. 3, https://www.opic.gov/sites/default/files/files/2015%20Financials_022416-508final.pdf

52. Overseas Private Investment Corporation, Telephone Interview with Senior Official at OPIC on September 24, 2014.

53. Ibid.

54. Ibid.

55. MIGA, Telephone Interview with Senior Official at MIGA on September 22, 2014.

56. MIGA: Oil and Gas, "Securing Oil and Gas Investments," MIGA Briefs, April 2013, http://www.miga.org/documents/oil&gasbrief.pdf

57. Berne Union Statistics 2011–2015, September 2016, p. 22, http://www.berneunion.org/wp-content/uploads/2016/09/Berne-Union-2016-Charts-and-numbers-for-website.pdf

58. Bloomberg New Energy Finance, "Fossil Fuel Divestment: A $5 Trillion Challenge," August 26, 2014, http://about.bnef.com/white-papers/fossil-fuel-divestment-5-trillion-challenge/

59. Ibid.

60. Gary C. Biddle, Gilles Hilary, and Rodrigo S. Verdi, "How Does Financial Reporting Quality Relate to Investment Efficiency?" *Journal of Accounting and Economics,* Vol. 48, No. 2–3 (2009): 112–131; R. Bushman and A. Smith, "Financial Accounting Information and Corporate Governance," *Journal of Accounting and Economics* 31, (2001): 237–333; P. Healy and K. Palepu, "Information Asymmetry, Corporate Disclosure, and the Capital Markets: A Review of the Empirical Disclosure Literature," *Journal of Accounting and Economics*, Vol. 31 (2001): 405–440; R. Lambert, C. Leuz, and R. Verrecchia, 2007. Accounting Information, Disclosure, and the Cost of Capital," *Journal of Accounting Research,* Vol. 45, p. 385–420; Biddle, G. and G. Hilary. "Accounting Quality and Firm-level Capital Investment," *The Accounting Review*, Vol. 81 (2006), 963–982.

61. Nick A. Owen, Oliver R. Inderwildi and David A. King, "The Status of Conventional World Oil Reserves—Hype or Cause for Concern?" *Energy Policy*, Vol. 38, No. 8 (August 2010), p. 4743–4749.

62. Ibid.

63. Ibid.

64. DOE, Telephone Interview with a Senior Official in the Office of International Affairs at the U.S. Department of Energy on September 29, 2014.

65. BP, "BP Statistical Review of World Energy 2014," http://www.bp.com/statisticalreview

66. Conglin Xu and Laura Bell, "Worldwide Reserves, Oil Production Post Modest Rise," *Oil & Gas Journal,* 12/02/2013, http://www.ogj.com/articles/print/volume-111/issue-12/special-report-worldwide-report/worldwide-reserves-oil-production-post-modest-rise.html

67. U.S. Energy Information Agency, Crude Oil Proved Reserves 2016, https://www.eia.gov/beta/international/data/browser/#/?tl_type=p&f=A&ct=0&c=ruvvvvvfvtvnvv1urvvvvfvvvvvvfvvvou20evvvvvvvvvnvvuvo&ord=SA&pa=000000000000000000008&tl_id=5-A&cy=2016&v=H&vo=0&so=0&io=0&start=1980&end=2016

68. Klaus Mohn and Petter Osmundsen, "Asymmetry and Uncertainty in Capital Formation: An Application to Oil Investment," *Applied Economics*, (June 2011): 4387–4401.

Chapter 5

Geopolitics, Multilateral Treaties, Forums, Diplomacy and National Security

The previous chapters have focused on the interaction between individual organizations and institutions. The first three sections in this chapter examine the interaction between organizations and institutions acting together, for example, multilateral treaties, international energy forums and international financial reporting institutions. The last three sections examine some unique relationships between IOCs and host governments, for example, signature bonuses, energy diplomacy and foreign direct investment.

MULTILATERAL TREATIES

In 2010, Goldthau observed that during the previous twenty-five years, the political and economic mainstream promoted the liberalization of energy trading, for example, the World Trade Organization, the North American Free Trade Agreement, the Energy Charter Treaty (ECT), the Asia-Pacific Economic Cooperation (APEC), and bi-regional forums such as the EU-Gulf Cooperation Council and the EU-Russian Dialogue, but direct foreign investments in energy were still managed through a "patchwork of national laws and bilateral treaties."[1] The Permanent Court of Arbitration only lists two multilateral investment treaties, the Convention on the Settlement of Investment Disputes between States and Nationals of Other States (1966) and the Energy Charter Treaty (1994). There are still no internationally agreed upon set of rules for trading energy resources and energy investments. The absence of an international agreement or even a substantial number of multilateral agreements is reflected in the small number of cases that have been brought before the ICSID and other tribunals, in which the legal instrument being referenced was a multilateral treaty. For example, of the sixty-three investment cases related to oil and gas reported by UNCTAD, forty-four were being arbitrated pursuant to a bilateral investment treaty between the host country and the home country of the investor; thirteen were being arbitrated pursuant to the Energy Charter Treaty (a multilateral trade agreement); three pursuant to the provisions of the North Atlantic Free Trade Agreement and three pursuant to a bilateral investment treaty and the Energy Charter Treaty.

Does the absence of globally agreed upon energy investment rules or even a substantial number of multilateral agreements matter? In chapter 3, it was noted that in 2014 and 2015, the total number of new deals (royalty/tax agreements, PSAs, service contracts, JVs, acquisitions and divestitures) in the exploration and development segment of the oil and gas industry, were 1,467 and 910, respectively. However, only two cases in 2014 and nine cases in 2015 were filed in an international court or tribunal.[2] Therefore, compared to the number of new deals made in 2014 and 2015, a relatively small number of transactions resulted in a dispute that had to be adjudicated by reference to a bilateral or a multilateral treaty. It was also noted that the outcome of an international court proceeding or tribunal is still dependent on the standard of treatment applied by the court or tribunal and the definition of fair and equitable compensation. Therefore, even if there was an international energy investment treaty or an extensive network of multilateral energy investment treaties, the problems of interpretation of specific provisions and standards of treatment would still exist.

INTERNATIONAL ENERGY FORUMS

Although no set of globally agreed upon energy investment rules or even a substantial number of multilateral agreements have been created, the oil and gas community comprised of national governments, NOCs and IOCs have created some institutions for exchanging information.

The International Energy Forum

The International Energy Forum (IEF) met for the first time in Paris in 1991. Its biennial ministerial meetings are the largest gathering of energy ministers in the world. Through the Forum and its associated events, IEF ministers, their officials, energy industry executives and other experts exchange information regarding common energy interests and global energy security. The IEF's mission is "to foster greater mutual understanding and awareness of common energy interests among its members."[3] The IEF has seventy-six member countries and all of them have signed the IEF Charter, which outlines a framework for "global energy dialogue through this intergovernmental arrangement,"[4] The member countries of the IEF account for approximately 90 percent of world oil and gas supply and demand. Its members include not only the oil-consuming and oil-producing countries of the IEA and OPEC, but also countries outside of their memberships, including Argentina, China, India, Mexico, Oman, Russia and South Africa.

However, this inclusiveness has not guaranteed a constructive dialogue.[5] For example, during a meeting of the IEF in London in 2008, it was suggested that planned capital expenditures be collected from IOCs and NOCs and be published. However, persuading IOCs and NOCs to publish their planned capital expenditures proved to be impossible because it demanded a high level of accuracy and required companies to disclose proprietary information about their current and future operations.[6] Consequently, although transparency would have increased market efficiency and reduced risk, IOCs and NOCs still only report at the level of detail required in their home country. Consequently, in the last twenty-five years, the main achievement of the IEF has been to raise awareness of the high level of energy interdependence among nations and the compilation and publication of historical data regarding production, consumption, and capacity expansion, but not future investment spending (CAPEX). Three of these IEF initiatives are discussed below.

Joint Oil Data Initiative (JODI)

At the 7th IEF Ministerial in Riyadh, Saudi Arabia in 2000, six organizations: Asia-Pacific Economic Cooperation (APEC), Statistical Office of the European Union (Eurostat), IEA, Latin American Energy Organization (OLADE), OPEC, and United Nations Statistical Division (UNSD), collaborated in the development of the *Joint Oil Data Exercise*. The *Joint Oil Data Exercise* was subsequently renamed the *Joint Oil Data Initiative* and was established as a permanent organization of the IEF. In January 2005, following the endorsement by energy ministers, the IEF secretariat assumed responsibility for coordinating the *Joint Oil Data Initiative*. The Initiative's oil database (JODI Oil) provides comprehensive, monthly statistics on global oil production and consumption.[7]

In November 2005, the JODI partners published the JODI Oil World Database (www.jodidata.org), which was one step to improve transparency in the oil and gas markets. The successful establishment of the oil database subsequently led to an initiative to develop a natural gas database (JODI Gas) and annual data on upstream and downstream capacity and expansion plans (JODI Investment).

FINANCIAL REPORTING

The institutions that provide the rules and improve the quality of financial reporting have not developed in response to the needs of the international oil and gas industry, but in response to international business in general. The institutions responsible for establishing rules for financial reporting include the International Accounting Standards Board (IASB), the Financial Accounting Standard Board (FASB), the Securities and Exchange Commission (SEC) and its counterparts in Europe and Asia. Beginning in the 1990s, the most significant developments in financial reporting have been, (1) a convergence between the International Financial Reporting Standards (IFRS) developed and updated by the IASB and the United States' Generally Accepted Accounting Principles (U.S. GAAP) developed and updated by the FASB and (2) the increasing adoption of IFRS worldwide. Furthermore, the influence of the IASB dominates other international institutions engaged in international accounting issues, such as the OECD Working Group on Accounting Standards, the UN Intergovernmental Working Group of Experts on International Standards of Accounting and Reporting and the European Union's Accounting Advisory Forum.[8]

The widespread adoption of IFRS since the 1990s has had a significant impact on financial reporting, capital markets, companies listed on public exchanges and investors. Companies listed on public exchanges can raise financial capital in other countries without having to first restate their financial results using that country's accounting standards. In addition, investors have access to financial data that is more consistent and comparable. In 2005, approximately 15,000 companies listed on exchanges around the world were preparing and presenting their financial results in compliance with the IFRS, and at the same time the differences between IFRS, U.S. GAAP and Japanese GAAP were also diminishing. Their convergence since 1990, is an example of how organizations, in this case investors, oil companies and financial organizations operating in cooperation with the institutions of the IASB, FASB and SEC, have reshaped the institutional environment to increase the efficiency of financial reporting and capital allocation.

SIGNATURE BONUSES

Signature bonuses are a part of the bid submitted by IOCs for the right to explore and develop a specific block or blocks in the host country's territory. These payments are made by the IOC to the host government immediately or shortly after a contract is signed. As the number of conventional drilling prospects has decreased, the value of these bonuses has increased. In the beginning, signature bonuses were between $1 million and $10 million per bid, however, recent exploration and development agreements have included signature bonuses exceeding $100 million.[9] The size of these signature bonuses and the occasional diversion of these payments by government officials to personal bank accounts have produced an ongoing controversy about their impact on host country governance.

For example, in July 2000, Marathon Oil made a signature bonus payment of $13.7 million to *Sonangol*, the NOC of Angola. It was the first in a series of three payments for the rights to develop a block offshore Angola.[10] The payment was made to a bank in Jersey, one of the Channel Islands off the United Kingdom. (Jersey laws permit a high level of bank secrecy and zero tax rates on income from foreign sources.) Subsequently, several news stories reported that the funds were rewired within a matter of hours to other *Sonangol* accounts around the world. Some of these were later proven to be the personal accounts of the Angolan president, Angolan government ministers and other government officials.[11]

Alleged corruption in other countries, similar to the kind reported in Angola, led to demands by several international organizations for transparency in the payments made by IOCs to national governments. In February 2001, in response to these demands, BP announced that it would begin publishing an annual statement of payments made to *Sonangol*. BP had paid $111 million for the development rights to Angola's Block 31.[12] *Sonangol* chief executive officer, Manuel Vicente, responded with a letter to BP's chief executive officer and chairman, John Browne, stating that *Sonangol* considered BP to be "violating the conditions of legal contracts signed with *Sonangol* . . . and if confirmed, is a sufficient reason to apply measures established in Article 40 of the PSA [production sharing agreement] i.e., contract termination."[13] Since that incident, no other major IOC has reported financial payments made in the form of signature bonuses. They have chosen instead, to conform to the host country's demand for confidentiality to preserve their relationship with the host government, avoid contract termination and maintain access to exploration blocks in future rounds of bidding.

The lack of transparency associated with these cash payments to governments and NOCs is considered to be a major cause of corruption in many oil-producing countries.[14] To address this problem, a public interest group, Publish What You Pay U.S., was formed in 2004 to advocate greater disclosure of payments made by international companies in general and IOCs in particular to host governments, however, the success of this campaign was limited for the reasons cited above. However, in October 2015, the U.S. Securities and Exchange Commission (SEC) was writing rules that would force U.S. companies to reveal what they pay foreign governments for oil, gas and other mineral exploration rights. These rules were being written at the same time a lawsuit was in progress, claiming the SEC was delaying implementation of the disclosures mandated by the U.S. Congress. Oil companies and trade groups objected to making so many details public, claiming that the SEC's initial rule would have given competitors an unfair advantage, and would violate laws of some host countries against such disclosures.[15]

However, on June 27, 2016, the SEC issued a rule requiring oil, gas and mining companies listed on the U.S. stock exchanges to disclose the payments they make to the United States and foreign governments.[16] How far other countries will go in establishing similar disclosure requirements is unclear.

ENERGY DIPLOMACY

Energy diplomacy can be defined as the use of foreign policy to secure access to energy supplies abroad and promote intergovernment cooperation in the energy sector.[17] Energy diplomacy is motivated by the belief that importing and exporting countries can enhance the security of supply, and improve and protect the competitive position of other sectors of their economy by giving their domestic oil and gas companies a competitive advantage in buying and selling oil and natural gas in the global market. For example, the Chinese government, through its state-owned banks and sovereign wealth fund, has been particularly aggressive in its lending to developing countries in Africa and Latin America. China has provided economic development assistance to these countries in exchange for access to natural resources. Countries that do not have credit ratings of sufficient quality to borrow from international commercial banks have gained access to capital by pledging proved oil and natural gas reserves in their country as collateral, in exchange for loans needed to develop their oil and gas resources and to fund public infrastructure projects. These agreements are frequently referred to as "loans for oil agreements."

Russia has, until recently, relied more on subsidies, price incentives and preferential relationships with the states of the former Soviet Union, and occasional military cooperation or intimidation. This system of barter was tried in Ukraine (in-kind gas payments in exchange for political allegiance to Russia) and Belarus (transit and political allegiance) for oil and gas.[18] In 2014, Russia began reducing subsidies and instead agreed to provide oil and gas under long-term supply agreements in exchange for the loans it needs to build pipeline and transport infrastructure and acquire entry into the downstream markets of other countries, particularly in Central and East Asia (Kazakhstan, Turkmenistan and China).

Loans for Oil Agreements

Pipeline projects are particularly well suited to loans for oil deals because after they are constructed, pipelines do not have alternative uses, thereby reducing the risk that either party will abrogate the contract (see for example the loans for oil and gas agreements between China and Russia, and China and Kazakhstan [Appendix VI]). In addition, in a loan-for-oil-deal, the resource owner retains ownership of the oil and natural gas, precluding the risk of expropriation.

Some analysts have dismissed loans for oil deals, asserting that they have not delivered what they promised. For example, despite its aggressive use of loans for oil, the share of equity oil (the amount of crude oil produced by China's foreign oil assets) remains relatively small. Equity oil accounts for only 12 percent of China's oil consumption and only 1 percent of global oil production.[19] Other analysts have drawn similar conclusions, but for a different reason. They claim that China has received relatively little in the form of engineering and construction contracts from the countries to which it has loaned money and in some cases, has received less oil and natural gas than was agreed on, particularly from Venezuela.[20, 21]

Energy diplomacy in the form of loans for oil has also been criticized for its impact on the borrowing country. These loans for oil frequently require that 60 percent to 70 percent of the funds advanced by the Chinese government be spent on projects constructed by Chinese engineering and construction companies, which may ultimately prove to be "white elephants," that is, an asset which the state cannot sell, but whose operating and maintenance costs exceed its actual use or value. In addition, it has been alleged that too little is being spent on developing the technical competencies of the local workforce that will be needed to operate and maintain these facilities after they are completed, including those that are for oil and gas production.[22]

Furthermore, critic's note that the market for oil is global and "liquid" and argue that these characteristics limit the value of energy diplomacy. For example, they argue that an increase or decrease in the global supply of oil will immediately translate into a change in the price of oil worldwide for all oil-consuming countries. Consequently, it does not matter who gets the crude out of the ground (China, United States or Venezuela) but how much oil is available worldwide.[23] They also argue that energy diplomacy aimed at foreign suppliers makes no difference to producing countries, provided the crude oil they produce reaches the global market somehow.[24] However, this microeconomic analysis gives little, if any, weight to the fact that energy diplomacy involves more than simply establishing a diplomatic relationship between an oil-exporting country and an oil-importing country. Energy diplomacy frequently leads to legally binding contracts that include specific price provisions that may lock in a below-market oil or natural gas price and a dedicated source of supply for years. These benefits are directly related to the issues of organizational arrangements, gained control, asset specificity, strategic behavior, contractual safeguards, surrounding uncertainty and bargaining strength and therefore should not be dismissed as unimportant.

In addition, the market for natural gas is still a regional market, because most natural gas is delivered by fixed pipeline rather than by LNG tankers which can efficiently transport LNG worldwide. Even the neoclassical economists would acknowledge that in a regional market of natural gas pipelines, energy diplomacy can be effective, because a natural gas pipeline connects the exporting country's gas fields and the importing country's markets, and after the pipeline is built, there are no efficient alternatives for either the importer or the exporter. However, as already observed, oil and LNG can be transported by ocean tanker to distant customers at a reasonable cost, making them attractive for expropriation by host countries. This may explain why oil and LNG assets are expropriated more often and disputes between host governments and foreign investors arise more often in oil and LNG project than in pipeline projects. Of the sixty-three arbitration cases examined in this book, fifty-nine were related to "extraction of crude oil and natural gas" and only four were related to "transport via pipeline."

FOREIGN DIRECT INVESTMENT AND NATIONAL SECURITY

Every country has rules regarding foreign direct investment. These rules serve a variety of purposes, including economic stabilization, economic development, technology transfer and national security. In the United States, the Committee on Foreign Investment in the United States (CFIUS) is an interagency committee "authorized to review transactions that could result in control of a U.S. business by a foreign person ('covered transactions') to determine the effect of such transactions on the national security of the United States."[25] For example, in 2005, CFIUS was involved in the CNOOC's proposed acquisition of Unocal (a U.S. oil company). Although this acquisition was not opposed by CFIUS or the Bush Administration, it was criticized by several members of Congress. After a vote in the United States House of Representatives, the bid was referred to the president on the premise that its implications for national security needed to be reviewed. While this review was still in progress, Unocal was acquired by Chevron for $18 billion, making the issue moot.[26]

A similar situation occurred in 2006, when the state-owned Dubai company, DP World, sought to manage some terminal operations at six American ports, but subsequently dropped out of the deal, in response to bipartisan opposition in the U.S. Congress. DP World, stated that at the direction of Dubai's ruler it would "transfer" to an American company the leases to manage some of the busiest terminals in the United States, including some in New York, Newark, Baltimore and Miami. The decision by the Dubai government avoided a confrontation between the Congress and President Bush, which the president would have most likely lost, given that the House Appropriations Committee had voted 62 to 2 to reject the deal, because it allowed the sale of some terminal operations to an Arab state company.[27] Similarly, in Russia and China, foreign direct investment rules require, among other things, that all foreign investors have Chinese or Russian partners.

NOTES

1. Andreas Goldthau and Jan Martin Witte, editors, *Global Energy Governance–The New Rules of the Game.* (Berlin: Global Public Policy Institute, 2010), p. 13.
2. EY, Global Oil and Gas Transactions Review 2015, http://www.ey.com/Publication/vwLUAssets/EY-global-oil-and-gas-transactions-review-2015/$FILE/EY-global-oil-and-gas-transactions-review-2015.pdf
3. International Energy Forum, http://www.ief.org/about-ief/history.aspx
4. Ibid.
5. Ibid.
6. Enno Harks, *The International Energy Forum and the Mitigation of Oil Market Risks*, Chapter 12 in Andreas Goldthau and Jan Martin Witte, editors, *Global Energy Governance—The New Rules of the Game.* (Berlin: Brookings Institution Press, 2010), p. 261.
7. Joint Organizations Data Initiative website, http://www.jodidata.org/about-Jodi.aspx
8. Jochen Zimmermann and Jorg R. Werner, *Regulating Capitalism? The Evolution of Transnational Accounting Governance.* (New York, NY: Palgrave McMillan, 2013), p. 15.
9. Andrew Inkpen and Michael H. Moffett, *The Global Oil and Gas Industry—Management, Strategy and Finance.* (Tulsa, OK: PennWell Corporation, 2011), p. 228.
10. David Leigh, "Angolan Oil Millions Paid into Jersey Accounts," *The Guardian,* November 4, 2002; Justin Pearce, "IMF: Angola's 'missing millions': The Sum is Three Times the Amount Paid in Aid," BBC, Friday October 18, 2002; "Angola Battles with IMF Accusation of Corruption," *Alexander's Gas and Oil Connections,* News and Trends Africa, Vol. 7, No. 22, November 13, 2002.
11. Andrew Inkpen and Michael H. Moffett, *The Global Oil and Gas Industry—Management, Strategy and Finance.* (Tulsa, OK: PennWell Corporation, 2011), p. 229.
12. Ibid.
13. Ibid.
14. Ibid.

15. Jennifer A. Dlhouhy, "SEC Plans Oil Payment Transparency Rule by June 2016," Fuelfix, October 5, 2015, http://fuelfix.com/blog/2015/10/05/sec-plans-oil-payment-transparency-rule-by-june-2016/

16. Marco Simons, "SEC Issues Long-Awaited Transparency Rule for Oil, Gas and Mining," EarthRights International, June 28, 2016, https://www.earthrights.org/media/sec-issues-long-awaited-transparency-rule-oil-gas-and-mining

17. Andreas Goldthau, and Jan Martin Witte, *Global Energy Governance—The New Rules of the Game*, Chapter 2—Energy Diplomacy in Trade and Investment of Oil and Gas. (Berlin: Global Public Policy Institute, 2010).

18. Ibid.

19. Kang Wu, "China's Overseas Oil and Gas Investment: Motivation, Strategies, and Global Impact," *Oil and Gas Energy Law,* Vol. 6, No. 1 (2008), p. 1–9.

20. Kelly Hearn, "Venezuelan Oil a Risky Investment for China," *The Washington Times*, Monday, March 12, 2012, http://www.washingtontimes.com/news/2012/mar/12/venezuelan-oil-a-risky-investment-for-china/#ixzz31tLhnzrI

21. Bo Kong, *China's International Petroleum Policy.* (Santa Barbara: Praeger Security International, 2010), p. 137.

22. Andrew Inkpen and Michael H. Moffett, *The Global Oil and Gas Industry—Management, Strategy and Finance.* (Tulsa, OK: PennWell Corporation, 2011).

23. Andreas Goldthau and Jan Martin Witte, *Global Energy Governance—The New Rules of the Game*, Chapter 2—Energy Diplomacy in Trade and Investment of Oil and Gas. (Berlin: Global Public Policy Institute, 2010).

24. Ibid.

25. CFIUS, Committee on Foreign Investment in the United States, http://www.treasury.gov/resource-center/international/Pages/Committee-on-Foreign-Investment-in-US.aspx

26. Associated Press, "China's CNOOC Drops Bid for Unocal," August 8, 20015, http://www.nbcnews.com/id/8795682/ns/business-oil_and_energy/t/chinas-cnooc-drops-bid-unocal/

27. David E. Sanger, "Under Pressure, Dubai Company Drops Port Deal," New York Times, March 10, 2016.

Chapter 6

Case Studies

This chapter includes six case studies. These cases were chosen because of their size and complexity and because together they illustrate the wide range of risks that can be encountered during oil and gas exploration projects. The events in each case study are presented in chronological order. At the end of each case there is an analysis of those events as they relate to subjects discussed in the previous chapters. These cases include (1) the Frade oil field offshore Rio de Janeiro, Brazil; (2) the Gorgon Natural Gas project in the Indian Ocean off the northwest coast of Australia; (3) the Baku-Tbilisi-Ceyhan (BTC) pipeline through Azerbaijan, Georgia and Turkey; (4) the Deepwater Horizon oil drilling rig in the Gulf of Mexico; (5) Expropriation of Oil and Gas Assets in Venezuela and Argentina; (6) and the TNK-BP joint venture in Russia.

CASE 1—THE FRADE FIELD

The Frade oil field is in the Northern Campos Basin, approximately 370 kilometers offshore Rio de Janeiro, Brazil, at a water depth of 1,100 meters (3,609 feet). The field was discovered in 1986. The project is a JV in which Chevron is the project operator holding a 51.7 percent interest. Petrobras has a 30 percent interest and Inpex/Sojitz has an 18.3 percent interest.[1]

Conceptual engineering studies and the acquisition of 3-D seismic data were completed by Texaco in 2000.[2] In June 2008, the process of installing subsea pumps and valves on the seafloor began. There were several delays related to the drilling rig and the floating, production, storage and offloading vessel (FPSO).[3] Recoverable reserves are estimated at 200 to 300 million barrels of oil. The cost of developing the field was estimated at $2.8 billion.[4]

Oil production began on June 20, 2009. The project was expected to achieve peak production of 90,000 barrels per day of crude oil and petroleum gas liquids in 2011.[5]

Timeline of Events

On November 7, 2011, oil began to seep from the cracks in the seabed above the field after a drilling accident.[6]

On November 9, 2011, the wellhead was tightly sealed and the leak was stopped in less than four days. Approximately eighteen vessels were deployed to support the oil sheen remediation and the activities necessary for well-abandonment. The total oil spilled was estimated at 2,700–3,600 barrels.[7]

On November 20, 2011, development drilling operations in the field remained suspended, but production from the field was maintained at approximately 79,000 barrels of oil equivalent per day.[8]

On November 22, 2011, the Brazilian government imposed a $28 million fine on Chevron for causing an offshore oil spill.[9] The Brazilian National Petroleum Agency (ANP) cited Chevron for twenty-five infractions related to the spill. Chevron agreed to pay the agency more than $17 million in fines. Transocean, the drill rig operator, was cleared of any liability.[10]

On November 25, 2011, Brazilian prosecutors sought 20 billion rials (US$10.9 billion) in damages from Chevron and offshore drilling contractor Transocean Ltd for the 3,600 barrels leaked from the Frade field.[11]

On December 2, 2011, Chevron Brazil Upstream Frade, a subsidiary of Chevron Corp., was ordered by ANP to shut down one of its eleven production wells and four "produced water" injection wells at Chevron's offshore Frade FPSO facility. The production wells cited by the ANP accounted for less than 10 percent of Frade's total production of about 79,000 barrels per day.[12]

The possible existence of a hydrogen sulfide (H_2S) gas leak was the basis for ANP requesting that the wells be closed temporarily. Chevron said that it respected the decisions of the ANP and would respond appropriately to the agency's requests for additional information. Chevron also said it would continue to keep the Brazilian agencies fully informed and work with them to address their specific concerns regarding Chevron's activities in Brazil. Chevron informed the ANP that it had conducted regular monitoring of the H_2S leak and had safety systems and processes in place to ensure the safety of employees, contractors and operations. Chevron said it was confident it would successfully respond to the ANP's concerns and resume operation of its production and injection wells.[13]

On February 3, 2012, Chevron cemented the well to stop the oil and gas seep through the fissures as a first step toward the capping and abandonment of the appraisal well.[14] On March 3, 2012, Chevron identified a new small seep in another part of the Frade field. Chevron notified the Brazilian authorities and immediately placed containment devices on the source. The total volume of this intermittent seep was approximately one barrel of oil.[15]

On March 15, 2012, Chevron's subsidiary, Chevron Brasil Upstream Frade Limitada, requested authorization to suspend production operations at the Frade Field because of the new seep and subsidence in the area. The company said it would conduct a comprehensive technical study and prepare a complementary study to better understand the geological features of the area, by working with their partners and seeking necessary approvals from the ANP. The decision was also endorsed by Chevron's partners. Chevron filed its request with the appropriate Brazilian regulatory agencies and expected a timely response.[16]

However, on April 4, 2012, A Brazilian federal prosecutor initiated a second 20 billion rial ($10.9 billion) lawsuit against Chevron and driller Transocean, doubling the claim against the companies, even though critics said the prosecutor was overreaching.[17] Both companies also faced criminal and civil lawsuits related to the spill. The criminal lawsuit was dismissed, but federal prosecutors appealed that decision. The two civil lawsuits sought approximately $21.8 billion in damages. In response, Chevron offered to pay approximately $150 million to settle the case.[18]

On August 6, 2012, Chevron appealed against an injunction banning it and its drilling contractor Transocean Ltd from operating in Brazil while civil and criminal charges over the oil spills were being adjudicated.[19]

However, on August 29, 2012, the injunction banning Chevron and Transocean from operating in Brazil was upheld by a panel of three Brazilian federal judges, while charges related to the November 2011 oil spill were being considered by the Brazilian courts.[20]

On February 20, 2013, a Brazilian judge dismissed criminal charges against Chevron Corp., Transocean Ltd and seventeen of their employees related to the offshore oil spill.[21]

On April 8, 2013, Chevron Corp was authorized to restart production at the Frade field for a period of twelve months. Output from an additional two wells was also approved, but only for two months.[22] On April 15, 2013, final documents were issued allowing production to resume.[23] Production was therefore shutdown from approximately March 15, 2012 to April 15, 2013, forcing Chevron and its partners to absorb the fixed costs associated with the leasing and maintenance of the drilling rig and the FPSO vessel.

On October 1, 2013, a Brazilian federal judge dismissed the two lawsuits totaling $21.8 billion against Chevron and approved a negotiated settlement for an undisclosed amount, a decision that closed a two-year legal battle over the 3,600 barrels of oil spilled in the Frade Field.[24] On April 1, 2014, the ANP authorized Chevron to resume full production at the Frade field.[25]

Case Analysis

Estimates were that 2,400–3,600 barrels of crude oil was leaked into the Atlantic Ocean.[26] In contrast, the consensus estimate of the oil spill in Prince William Sound in 1989, when the Exxon Valdez ran aground, was 260,000 barrels [27] and the U.S. Government has estimated the oil spill in the Gulf of Mexico in 2010, when BP's Deepwater Horizon drilling platform caught fire and sank, at 4.9 million barrels of oil.[28]

Although, Chevron and its partners were the targets of a $21.8 billion lawsuit along with criminal charges for a relatively small oil spill, the criminal charges were ultimately dismissed by a Brazilian federal judge and the civil charges were settled for an undisclosed amount. The excessive charges originally brought against Chevron and its partners suggest that some prosecutors in Brazil may have been driven more by political considerations than legal or substantive considerations. This may explain in part why the Rule of Law in Brazil was rated 51.7 on a scale of 0 to 100.

In this case, however, the Brazilian federal judge's decision made it unnecessary for Chevron to bring the case before an international court or tribunal. A country's legal system therefore does not need to work perfectly, but only well enough to compensate for partiality in other parts of the country's legal system.

Second, Chevron's offer to pay more than $150 million to settle the case rather than keep production shutdown after the panel of three federal judges upheld the injunction banning Chevron and Transocean from operating in Brazil, demonstrates the bargaining leverage that national governments and their agencies acquire after the sunk costs in exploratory drilling, development and production equipment have been made.

Third, the defendants, Chevron and Transocean, did not bring the dispute with the Brazilian regulatory agencies or the prosecutor before an international court or tribunal, but instead chose to seek a solution in the Brazilian courts and ultimately agreed to a settlement. This supports the view that IOCs are reluctant to resort to an international court or tribunal, except in those cases in which the investor believes he is unlikely to receive a fair hearing in the host country's courts.

Fourth, there is no evidence that the U.S. government tried to intervene on Chevron's behalf, supporting Lipson's observation that national governments seldom intervene in investment disputes on behalf of their citizens.

CASE 2—THE GORGON FIELD

This case is included because of the size and complexity of the project, the magnitude of the cost overruns and because it illustrates the financial consequences of unanticipated economic events. The Gorgon gas field, in the Indian Ocean off the northwest coast of Australia was once considered not recoverable, so-called "stranded gas," because of its complex geology, the environmentally sensitive areas near it and the high exploration and development expenses associated with the project.[29] The Gorgon project is a JV between Chevron (47.3 percent, project operator), ExxonMobil (25 percent), Shell (25 percent), Osaka Gas (1.25 percent), Tokyo Gas (1 percent) and Chubu Electric Power (0.417 percent) [30] When the project began in 2009, the cost was estimated at 43 billion Australian dollars, (U.S. $37 billion; 3.4 trillion Japanese yen). The field is estimated to contain more than 13.8 trillion cubic feet of natural gas (the equivalent of 2.25 billion barrels of oil) and is expected to have a production life of approximately forty years.[31]

Timeline of Events

On April 24, 2009, Chevron requested approval from Australian authorities to expand its Gorgon LNG production capacity to fifteen million metric tons per year to compensate for rising development and production costs.[32] On April 30, 2009 the Western Australian Environmental Protection Agency approved Chevron's request for authorization to add a third five million metric-tons-per-year LNG train (unit) to the original two-train proposal already approved for Barrow Island. Chevron continued to assess the geologic and environmental conditions as it worked toward a final investment decision in the second half of 2009.[33]

On August 10, 2009, the environment minister Donna Faragher, gave final environmental approval for the proposed Gorgon gas field development on Barrow Island, but this approval also imposed higher levels of environmental protection for regionally significant coral reefs and flatback marine turtles.[34] On September 1, 2009, minister of mines and petroleum, Norman Moore, offered the Gorgon JV partners production licenses for its biggest natural gas fields.[35] On September 14, 2009, development proposals for the project were approved by the premier of Western Australian state, Colin Barnett, and production licenses were granted by the Australian minister for resources and energy, Martin Ferguson.[36]

On December 1, 2009, the construction of a three-train (unit) 15 million-metric tons-per-year LNG facility for the Gorgon LNG project began.[37] In the next several months Chevron made a succession of gas discoveries in the Gorgon field. On December 14, 2009, Chevron made a second natural gas discovery in the Carnarvon Basin offshore Western Australia with its Satyr-1 well.[38] On January 26, 2010, a third gas discovery was made in the Greater Gorgon Area in the Carnarvon Basin offshore Western Australia.[39]

On February 9, 2011, Chevron made an additional discovery in the Orthrus-2 well located in the WA-24-R permit area in the Carnarvon Basin offshore Western Australia.[40] On January 19, 2012, Chevron appraised the Satyr-3 well in the Exmouth Plateau area of the Carnarvon Basin, offshore Western Australia and concluded that it was commercially viable.[41]

On July 23, 2012, Chevron made a natural gas discovery in the Greater Gorgon area of the Carnarvon Basin, offshore Western Australia.[42] In September 2012, Chevron continued a successful run in the Greater Gorgon area offshore Western Australia with the announcement of positive results from the Satyr-2 exploration well.[43]

In October 2012, Australian energy minister, Martin Ferguson, said the carbon dioxide (CO_2) capture and sequestration project related to Chevron's Gorgon LNG project was on schedule and would begin injection of CO_2 in 2015. The world's largest carbon capture and sequestration project was designed to inject 3.5 million tons of CO_2 per year at a depth of about 2,300 meters below Barrow Island, off Western Australia.[44]

On December 6, 2012, Chevron increased the estimated cost of the Gorgon project by $15 billion, bringing the total cost of the project to $52 billion. Chevron attributed the cost increase primarily to labor shortages, logistics challenges and the strength of the Australian dollar. Despite the increase in the project's cost, Chevron said that the increase in the price of crude oil would still make the project profitable, because LNG prices usually move in tandem with the oil prices, which Chevron estimated had increased by 80 percent between December 2009 and December 2012. However, the increased cost of the project was expected to significantly reduce the project's profitability. Some analysts estimated that if the price of Brent Crude dropped below $80 per barrel, the project would not be competitive with other energy sources and would therefore be unprofitable.[45]

On December 12, 2013, Chevron announced that the estimated cost of the Gorgon LNG project had increased from $52 billion to $54 billion, bringing the total cost increases to $17 billion Chevron said that cost overruns and delays were the result of the high value of the Australian dollar, high Australian wages, low productivity, weather delays and the logistical challenges of building an LNG plant on Barrow Island (a Class A nature reserve). The secretary of the Western Australia Maritime Union of Australia (WAMUA), Christy Cain, claimed that "red tape" and waste were the reasons the cost of the project had increased. The Maritime Union also disputed Chevron's claims of low productivity among workers on the project and said workers were being unfairly blamed for the cost overruns. The Australian Mines and Metals Association (AMMA), a resources industry group, claimed that the high wages in Australia were a major factor contributing to the cost overruns.[46]

In February 2014, former labor resources minister and former union leader, Martin Ferguson, also blamed the WAMUA for the cost overruns and delays in the Gorgon project.[47] The WAMUA responded that the AMMA and Ferguson were blaming the WAMUA for problems on the Gorgon project, because the union had sought to negotiate a new Enterprise Bargaining Agreement for maritime workers working in the offshore oil and gas industry. WAMUA secretary, Christy Cain, said maritime workers were being used as scapegoats by Chevron Australia.[48]

On March 12, 2014, the Associated Press and the Dow Jones Newswire reported that the Gorgon project was 78 percent complete and that two-thirds of gas production was already committed to buyers. However, the start-up date for "first gas production" was revised from the middle of 2014 to the middle of 2015.[49] On August 1, 2014, Chevron reported that the Gorgon project was 83 percent complete and that "first gas production" was still expected in the middle of 2015.[50]

However, on August 25, 2014, Reuters reported that Chevron was finding it difficult to find buyers willing to sign twenty-year sales contracts for gas from its Gorgon LNG export plant in Australia and commented that the high level of unsold LNG expected from the Gorgon field demonstrated that the shale gas "boom" in the United States had significantly reduced the profitability of the Gorgon project and other major LNG investments in Australia.[51]

On October 11, 2016, the spot price of Brent Crude was $50.48 per barrel, well below the breakeven point of $80 per barrel referenced above.[52] However, the value of the Australian dollar had declined from a high of 0.917 Australian dollars per U.S. dollar in April 2011 to 1.313 Australian dollars per U.S dollar in October 2016, thereby mitigating the impact on the cost of paying local Australian providers of equipment, services and labor to the project.

Case Analysis

This project does not demonstrate political risk, but rather commercial risk. The increase in the cost of the project from $37 billion to $54 billion is principally the result of the (1) rapid increase in the value of the Australian dollar (as a result of the increase in the global demand for the iron ore, bauxite, coal and other raw materials Australia produces and sells to the rapidly growing economies of Asia) and (2) the approval of the third LNG unit being conditioned on the implementation of costlier environmental protection measures. The commercial risk associated with this project is the consequence of the significant decline in world oil and gas prices between April 2009 and July 11, 2016. These factors, however, lie outside the scope of this book.

Nevertheless, a project in which the original cost estimate was overrun by 17 billion U.S. dollars (46 percent) might be expected to give rise to considerable litigation, as each party tried to recoup its position under the provisions of a stabilization or equilibrium clause in one or more contracts. For example, the value of the Australian dollar increased from 1.3996 Australian dollars per U.S. dollar in April 2009 to 0.9328 Australian dollars per U.S. dollar in July 2011, which made it more expensive for Chevron and its subcontractors to pay for locally sourced equipment, services and labor.

Despite these adverse developments, neither Chevron and its partners nor any of the other parties to the many contracts in force during this project have sought compensation in the Australian courts or elsewhere. This is in part because the changed economic circumstances (the appreciation of the Australian dollar, the addition of a third LNG unit and the significant decrease in the price of oil and gas) were not the result of a change in government policy, but rather the consequence of natural market forces.

CASE 3—THE BAKU-TBILISI-CEYHAN PIPELINE

This case is included because of the large number of partners involved in the project, the social, political and environmental issues related to the project and its transit route through three developing countries, any one of which could have disrupted the construction or the operation of the pipeline. The BTC pipeline links Baku, Azerbaijan with the Turkish Mediterranean port city of Ceyhan. It is 1,100 miles long (1,768 km), requires more than 100 surface stations, associated pumping facilities and plants to maintain the flow of oil, and can carry one million barrels of oil per day. The pipeline passes through Azerbaijan, Georgia and Turkey.[53] The pipeline cost approximately $3.7 billion to build.[54]

The pipeline was constructed by a consortium of companies led by BP and was financed with a high level of debt using a project financing structure. The BTC Company shareholders include BP (30.1 percent), Azerbaijan (BTC) Limited (AzBTC, 25 percent), Chevron (8.9 percent), Statoil (8.71 percent), Türkiye Petrolleri Anonim Ortaklığı (TPAO, 6.53 percent), Eni (5 percent), Total (5 percent), Itochu (3.4 percent), Inpex (2.5 percent), ConocoPhillips (2.5 percent) and ONGC (Oil and Natural Gas Company Limited, 2.36 percent).[55] The pipeline was completed in May 2005.

Timeline of Events

In the spring of 1992, the Turkish prime minister, Suleiman Demirel, proposed to Central Asian countries including Azerbaijan, that the pipeline run through Turkey. Almost a year later (March 9, 1993) the first document related to the construction of the BTC pipeline was signed between Azerbaijan and Turkey, in Ankara, Turkey.[56] The Turkish route required that the pipeline from Azerbaijan run through Georgia or Armenia, but the route through Armenia was politically impossible because of the unresolved conflict between Armenia and Azerbaijan over the territorial status of Nagorno-Karabakh. This left the Azerbaijan-Georgia-Turkey route, which was longer and more expensive to build than a route through Armenia.[57] Discussions continued between the countries for more than five years before the project gained momentum.

On October 29, 1998, the Ankara Declaration was signed by the president of Azerbaijan, Heydar Aliyev; president of Georgia, Eduard Shevardnadze; president of Kazakhstan, Nursultan Nazarbayev; president of Turkey, Suleiman Demirel; and president of Uzbekistan, Islam Karimov.[58] Approximately a year later (November 18, 1999) the Inter-Governmental Agreement (IGA) in support of the pipeline was signed by Azerbaijan, Georgia, and Turkey during a meeting of the Organization for Security and Cooperation in Europe (OSCE) in Istanbul, Turkey.[59] "The IGA is a trilateral agreement between Azerbaijan, Turkey and Georgia, which affirms each country's support for the pipeline and documents several mutual undertakings to ensure the project's construction and operation. The IGA has the status of an international treaty."[60]

Almost three years later (August 1, 2002) the BTC pipeline company was established in London, making it possible to sign the Host Government Agreements (HGAs).

> "The HGAs are individual agreements between BTC Co and each of the countries [through which the pipeline passes]. They define the capital and resources that each signatory is to provide to the project, the timetable by which this project would be developed, the standards that it must meet, and the domestic legislation, both current and future, to which the project is subject. The HGAs have been incorporated into domestic law in all three host countries and overrides all domestic law (other than the national constitutions) where such law conflicts with the terms of the HGAs and the IGA. They are enforceable as contracts between BTC Co and each country."[61]

Each country signed a Joint Statement in which:

> "BTC Co and the host governments . . . committed to ensure that the pipeline project would adhere to the OECD Guidelines and the Voluntary Principles on Security and Human Rights. . . . [T]he Joint Statement is a Project Agreement under the HGA, making it a binding obligation on all parties. BTC Co has also covenanted to the project's financial backers to adhere to the Joint Statement."[62] Other legal agreements supporting this project include the Environmental and Social Impact Assessments, BTC Human Rights Undertaking and the Security Protocol.

Construction began in April 2003. The Azerbaijan section was constructed by Consolidated Contractors International of Greece, and the Georgia section was constructed by a JV of France's *Spie Capag* and UK Petrofac International. The Turkish section was constructed by BOTAŞ Petroleum Pipeline Corporation. Bechtel Corporation was the general contractor for engineering, procurement and construction.[63]

The financing for the project was agreed to in February 2004, after more than two years of appraisal of the potential environmental and social impacts of the project. Approximately 70 percent of the project costs were funded by a group of lenders including the IFC, the European Bank for Reconstruction and Development, the export credit agencies of seven countries and a syndicate of fifteen commercial banks.[64]

The first oil began to flow on May 10, 2005, and reached Ceyhan, Turkey on May 28, 2006.[65] In March 2009, the capacity of the BTC pipeline was increased to 1.2 million barrels per day.[66] From June 2006 to April 21, 2014, 2,500 tankers carrying 1.9 billion barrels of oil (256 million tons) delivered by the BTC pipeline had been shipped from the Turkish marine terminal in Ceyhan.[67] On August 11, 2014, the BTC Pipeline shipped its 2 billionth barrel of oil.[68]

Case Analysis

The design, construction and operation of the BTC pipeline had to overcome many political, social and environmental obstacles.[69] To overcome these obstacles, several of the strategies discussed in previous chapters were employed. These include (1) project financing using a high level of debt, (2) the participation of numerous public and private financial organizations, (3) numerous legal agreements designed to ensure continued cooperation between the parties in the consortium, (4) multilateral agreements between host governments that stipulated that the provisions in the principal legal agreement and other legal agreements override the domestic law of any of the countries involved in the project and (5) the inclusion of eleven equity partners. These arrangements reduced the risk of unilateral action by one of the sponsors or host governments that could impede the shipment of oil or result in expropriation of the fields or the pipeline. There has not been a major disruption or dispute since the project began operation in 2005, proving that multilateral agreements can be effective and that parties to such agreements generally honor their commitments.

CASE 4—THE DEEPWATER HORIZON DRILLING RIG

This case illustrates the operation of domestic environmental liability law as applied to an international corporation with its principal offices in another country. The Deepwater Horizon was an ultra-deepwater, offshore oil drilling rig owned by Transocean. The rig was built in 2001 in South Korea by Hyundai Heavy Industries, was commissioned by R&B Falcon (which later became part of Transocean), was registered in Majuro, Marshall Islands and was leased to BP from 2001 until September 2013. In September 2009, the rig drilled the deepest oil well in history at a vertical depth of 35,050 feet (10,683 meters), approximately 250 miles (400 km) southeast of Houston, in 4,132 feet (1,259 meters) of water.[70]

On April 20, 2010, while drilling at the Macondo Prospect, an explosion on the rig caused by a blowout, killed eleven crewmen and ignited a fire which was visible from thirty-five miles away (56 km). The fire could not be extinguished and, on April 22, 2010, the Deepwater Horizon sank, with the well discharging oil and natural gas at the seafloor for eighty-seven days, until it was successfully capped on July 15, 2010, making it the largest offshore oil spill in U.S. history.[71] The U.S. government estimated the total discharge of oil into the ocean at 4.9 million barrels (210 million U.S. gallons).[72]

Civil and criminal legal actions resulting from the loss of the Deepwater Horizon and the oil spill began almost immediately after the explosion, including many individual and class action lawsuits. Many claims were resolved administratively from a fund set up by BP for that purpose. The civil trial brought by the U.S. government began in early 2013 and was split into three phases, one to assign blame for the oil spill, a second to determine how much oil was spilled and a final phase to determine penalties for BP and Anadarko Petroleum Corp., its partner in the failed well.[73]

Timeline of Events

On April 20, 2010, the Deepwater Horizon drilling rig exploded, sending oil, gas and concrete up the wellbore onto the deck of the drilling platform before catching fire. The explosion killed eleven platform workers and injured seventeen others; another ninety-eight survived without serious physical injury.[74] Two days later, on April 22, 2010, the Deepwater Horizon rig sank.[75]

On April 24, 2010, BP reported a leak of approximately 1,000 barrels per day (42,000 U.S. gallons per day).[76] On April 26, 2010, an oil slick was reported 36 miles (58 km) southeast of Louisiana. Booms were set up to keep the oil from washing ashore[77] but by April 27, 2010, the oil slick had expanded 100 miles (160 km) across and was 20 miles (32 km) from the Louisiana coast.

On April 28, 2010, the National Oceanic and Atmospheric Administration estimated that the leak was likely to be 5,000 barrels per day (210,000 U.S. gallons per day), five times larger than initially estimated by BP.[78, 79] On May 13, 2010, Tony Hayward, the CEO of BP called the oil spill "relatively tiny" in comparison with the size of the "ocean."[80] On the same day, Transocean (a Swiss company and owner of the Deepwater Horizon drilling rig) filed to limit its liability under the Limitation of Ship Owner's Liability Act in the U.S. district court for the southern district of Texas to only its interest in the Deepwater Horizon, which it valued at $26.8 million.[81]

On May 19, 2010, oil washed ashore on mainland Louisiana.[82] A week later, on May 27, 2010, (1) President Obama announced a six-month moratorium on new deepwater oil drilling (500 feet [150 meters] of water or more);[83, 84] (2) BP Plc and Transocean Ltd. faced thirty-six lawsuits, including group cases with potentially thousands of plaintiffs, over environmental damage and personal injuries caused by the oil spill.[85]

On June 1, 2010, oil washed up on the beaches of Gulf Islands National Seashore.[86] On June 4, 2010, Tar balls arrived on beaches in Pensacola, Florida.[87] On June 11, 2010, Flow Rate Technical Group, a consulting firm, estimated that the leak might be 20,000 to 40,000 barrels per day (840,000 to 1,680,000 U.S. gallons per day.)[88]

On June 16, 2010, President Obama met with BP executives, during which BP agreed to fund a U.S. $20 billion escrow account administered by Kenneth Feinberg.[89] On June 23, 2010, oil appeared on Pensacola Beach and in Gulf Islands National Seashore, and officials warned against swimming for 33 miles (53 km) east of the Alabama line.[90]

On August 10, 2010, seventy-seven cases, including those brought by state governments, individuals, and companies, in the U.S. district court for the Eastern district of Louisiana were combined under Multidistrict litigation docket MDL No. 2179, captioned *In re: Oil Spill by the Oil Rig "Deepwater Horizon" in the Gulf of Mexico, on April 20, 2010*, presided over by U.S. district judge Carl Barbier.[91, 92]

On September 19, 2010, BP officially declared the oil well completely and permanently sealed.[93]

On December 15, 2010, the U.S. federal government sued BP Exploration and Production, Inc., and eight other corporations for unlimited liability, to pay the expenses involved in the cleanup and environmental recovery from the spill. The U.S. Department of Justice also filed a civil suit against BP and other defendants for violations of the Clean Water Act in the U.S. district court for the Eastern district of Louisiana; this case was subsequently consolidated with the other cases,[94] which were captioned *United States of America v. BP Exploration & Production Inc. et al., Civ. Action No. 2:10-cv-04536*.[95]

On April 21, 2011, BP filed $40 billion in lawsuits against rig owner Transocean, cement contractor Halliburton and blowout preventer manufacturer Cameron International Corporation. BP alleged that failed safety systems and irresponsible behavior of the contractors had led to the explosion.[96]

On March 2, 2012, BP settled with most private plaintiffs, BP initially estimated the cost of this settlement at $7.8 billion. In a subsequent regulatory filing, the cost of the settlement with private plaintiffs would increase to $9.2 billion.[97] On August 13, 2012, BP asked U.S. district judge Carl Barbier to approve this settlement, claiming that its actions "did not constitute gross negligence or willful misconduct." [98, 99] (Under the Oil Pollution Act of 1990, a company is only liable for $75 million in economic damages, provided it did not exhibit "gross negligence"; the U.S. federal government is required by law to pay the next $1 billion in claims out of the Oil Spill Liability Trust Fund discussed earlier.)[100]

On August 31, 2012, the U.S. Department of Justice, in response to BP's filing; and to ensure that BP could not use its filing and any possible acceptance of the settlement to escape a judgment of gross negligence in its own suits against BP,[101] filed papers describing the spill as an example of "gross negligence and willful misconduct."[102, 103] (A ruling of gross negligence would result in a fourfold increase in the penalties under the Clean Water Act and would increase damages in the other suits as well.)[104]

On November 14, 2012, BP agreed to pay $4.5 billion in fines and other payments, the largest of its kind in U.S. history. BP also agreed to plead guilty to eleven felony counts related to the deaths of the eleven workers.[105] [106] The Justice Department had previously filed criminal charges against one BP employee in April 2012 and against three BP employees in November 2012.[107, 108] Two of the employees were indicted on manslaughter charges for acting negligently in their supervision of key safety tests performed on the rig prior to the explosion and their failure to alert onshore engineers of problems with the drilling operation;[109] and two employees were charged with obstruction of justice and for lying to federal investigators, one was later found guilty. [110, 111, 112]

On January 3, 2013, The U.S. Justice of Department announced that "Transocean Deepwater, Inc. agreed to plead guilty to violating the Clean Water Act and would pay a total of $1.4 billion in civil and criminal fines and penalties"[113]

On January 13, 2013, Judge Barbier approved the medical benefits portion of BP's proposed $7.8 billion partial settlement (subsequently revised to $9.2 billion). People living at least for sixty days along the affected shores or were involved in the cleanup and who could document one or more specific health conditions caused by the oil or dispersants, were eligible for benefits. The settlement was intended to resolve more than 100,000 claims, which BP had originally estimated could cost about $7.8 billion. However, there was no limit to payouts under the medical settlement, which also included provisions allowing individuals to file for benefits in the future, if they could show that their illnesses were caused by the spill.[114]

On February 25, 2013, BP and its partners (Transocean and Halliburton) went on trial in the U. S. district court for the Eastern district of Louisiana in New Orleans to determine payouts and fines under the Clean Water Act and the Natural Resources Damage Assessment. The trial's first phase was to determine the liability of BP, Transocean, Halliburton and other companies, and to determine whether the companies had acted with gross negligence and willful misconduct.[115, 116]

On July 9, 2013, BP reported that it had spent approximately $25 billion on the Gulf Oil Spill, however, this did not include the $4.5 billion that was owed to the U.S. Government as the result of a settlement between BP and the U.S. Department of Justice for fines and penalties for violation of the Clean Water Act (see entry on November 14, 2012 above.)[117]

On September 30, 2013, the second phase of the trial of BP and its partners focused on how much oil spilled into the Gulf of Mexico and who was responsible for stopping it.[118] Claims against BP's drilling fluids contractor M-I LLC were dismissed by U.S. district judge Carl Barbier during the second phase of the trial. The judge also ruled out punitive damages against Cameron International, the manufacturer of the blowout preventer on the Deepwater Horizon rig.[119]

On February 12, 2013, the payment of claims by BP's claims administrator was placed on hold after BP filed an appeal with the 5th Circuit of Appeals regarding the $9.2 billion settlement BP had signed in March 2012.[120]

On November 25, 2013, BP argued before the Texas Supreme Court that it should be covered by a $750 million insurance policy taken out by Transocean Ltd., the Deepwater Horizon rig's owner, prior to the spill.[121] By December 17, 2013, BP had paid almost $13 billion in claims to businesses, individuals and federal and state governments.[122] By February 9, 2014, BP had sold $38 billion in assets, including half of its offshore platforms and refineries, to pay a projected total of $42 billion for the cleanup, compensation to businesses and individuals and other costs associated with the spill.[123]

On March 4, 2014, BP was told by a panel of the U.S. 5th Circuit Court of Appeals that it must abide by the terms of a $9.2 billion settlement with victims of the Deepwater Horizon oil spill. (BP failed to satisfy the judges that a claims administrator was misinterpreting the agreement and was paying claims that BP described as "fictitious" and "absurd.)"[124]

On April 20, 2014, BP refused to pay a $148 million invoice for U.S. government studies related to the 2010 oil spill, including research into its impact on dolphins, whales and oysters; and published figures indicated that BP had reduced its spending on the Natural Resource Damage Assessment (NRDA). Still, at the end of 2012, BP had paid $973 million for NRDA studies, and by April 2014 had paid $1 billion pursuant to the NRDA.[125]

On May 19, 2014, the U.S. 5th Circuit Court of Appeals refused to reconsider its previous ruling that businesses did not have to prove they were directly harmed by BP's 2010 Deepwater Horizon oil spill to receive settlement payments.[126]

On June 9, 2014, the U.S. Supreme Court declined a request from BP to block payments to businesses while it appealed the district and appeals courts' interpretation of the $9.2 billion settlement. This decision upheld the ruling by the U.S. 5th Circuit Court of Appeals in New Orleans that under the terms of the settlement, businesses claiming damages from the oil spill did not have to prove direct harm.[127]

On September 2, 2014, Halliburton Co. announced it had reached a $1.1 billion agreement that it said would settle most of the class action claims asserted by plaintiffs against the company following the Deepwater Horizon oil spill in the Gulf of Mexico. The proposed settlement, which was still subject to the U.S. district court for Eastern Louisiana's approval and an agreed-upon participation level by current claimants, would be paid into a trust fund until all appeals had been resolved in three installments over the next two years. The company previously had taken a $1.3 billion contingency loss for the Deepwater Horizon lawsuits in multiple districts. Halliburton said it would withdraw the proposed settlement if not enough claimants agreed to accept it.[128]

Two days later, on September 4, 2014, District Judge Barbier ruled that BP was guilty of "gross negligence and willful misconduct" under the Clean Water Act. He described BP's actions as "reckless", and Transocean's and Halliburton's actions as "negligent." He apportioned 67 percent of the blame for the spill to BP, 30 percent to Transocean, and 3 percent to Halliburton. Fines would be apportioned commensurate with the degree of negligence of the parties and the number of barrels of oil spilled. (under the Clean Water Act fines can be based on a cost per barrel of up to $4,300 per barrel, if in the judge's opinion the defendant's behavior constitutes "gross negligence.") The number of barrels was still in dispute at the conclusion of the trial. BP argued that 2.5 million barrels were spilled over the 87 days, but the U.S. government and the court contended that 4.2 million barrels were spilled.[129] The court's decision regarding "gross negligence and willful misconduct" could result in fines and penalties up to $18.1 billion (4,200,000 barrels x $4,300 per barrel = $18.1 billion) for violating the Clean Water Act.[130] This ruling would increase BP's total cost for the Deepwater Horizon oil spill well beyond the $43 billion it had already provided for.[131] BP issued a statement strongly disagreeing with the court's ruling and immediately filed an appeal.

On October 3, 2014, BP asked Judge Carl Barbier for a new judgment or retrial in the "gross negligence" suit, claiming that the verdict allegedly relied on evidence that had been excluded from court.[132]

On December 8, 2014, the U.S. Supreme Court refused to hear BP's appeal that some businesses and individuals were receiving payouts despite being unable to trace their losses to the effects of the BP oil spill, claims that BP had characterized as "fictitious and absurd". The U.S. Supreme court's refusal to hear BP's appeal meant that BP would have to make the payments.[133]

On February 13, 2015, the Texas Supreme Court ruled that BP was not owed coverage under Transocean's $750 million insurance policy because the court incorporated limitations on Transocean's underwater pollution liability from the BP-Transocean drilling contract into the insurance policy, thereby releasing Transocean and its insurers from paying BP's underwater pollution costs.[134]

On February 19, 2015, U.S. district judge Carl Barbier ruled that BP could pay a maximum civil penalty of up to $4,300 for each barrel of oil spilled under the Clean Water Act. He rejected BP's argument that the fines should be capped at $9.57 billion, approximately one-third lower than the $13.7 billion penalty federal prosecutors were seeking.[135]

On October 5, 2015, the U.S. Justice Department joined five Gulf states (Louisiana, Alabama, Florida, Mississippi and Texas) in announcing a settlement to resolve civil claims against BP arising from the Deepwater Horizon oil spill. This settlement resolved the governments' civil claims under the Clean Water Act and natural resources damage claims under the Oil Pollution Act, as well as economic damage claims of the five Gulf states and local governments. The resolution of these civil claims totaled $20.8 billion, and is the largest settlement with a single entity in the history of the U.S. Justice Department.

Under the terms of the consent decree lodged in federal court in New Orleans, BP agreed to pay the following:

- $5.5 billion penalty for violation of the federal Clean Water Act, plus interest, 80 percent of which is for restoration efforts in the Gulf region pursuant to a Deepwater-specific statute, the RESTORE Act. This is the largest civil penalty in the history of environmental law.
- $8.1 billion damages to natural resources in the Gulf, this includes $1 billion BP already committed to pay for early restoration, for joint use by the federal and state trustees in restoring injured resources.
- BP also agreed to pay up to an additional $700 million, some of which is for accrued interest and some specifically to address any subsequent natural resource problems that were unknown at the time of the agreement.
- BP will also pay $600 million for other claims, including claims for reimbursement for the costs of an assessment of federal and state natural resource damage, other unreimbursed federal expenses, and to resolve a False Claims Act investigation resulting from the spill.
- BP has also signed separate agreements to pay $4.9 billion to the five Gulf States and up to $1 billion to several hundred local governmental bodies to settle claims for economic damages they claim to have suffered as a result of the spill.[136]

On July 14, 2016, BP increased its total cumulative pretax charge for the Deepwater Horizon oil spill by $5.2 billion to $61.6 billion after making "significant progress in resolving outstanding claims." The $5.2 billion charge includes the estimated cost of settling all outstanding business and economic loss claims under the 2012 Plaintiffs' Steering Committee (PSC) settlement. Those are expected to be paid by 2019.[137]

Case Analysis

This case exhibits several of the features observed in previous chapters related to the management of risk in the international oil and gas industry. First, there is no evidence that the British government tried to intervene at any time on BP's behalf, supporting the observation that national governments do not intervene on behalf of investors involved in a dispute arising from foreign direct investment.

Second, international oil and gas companies are likely to use the domestic courts of the host country before resorting to international courts or tribunals, if the courts are thought to be reliable and fair. (see actions initiated by BP on April 21, 2011; March 2, 2012; August 13, 2012; October 2013; November 25, 2013; April 20, 2014; June 9, 2014, September 4, 2014; October 3, 2014; October 5, 2015; and July 14, 2016). BP's decision to use the U.S. court system to protect its interests and minimize its liability was probably made easier by the fact that in the United States the Rule of Law is rated 91, indicating a high level of fairness and reliability. (The ICSID does not require that domestic court remedies be exhausted before filing a claim with it.)

Third, like most international oil and gas companies BP chose to self-insure against expropriation and liability rather than purchase insurance against the potential risk of either. BP had some insurance through Lloyd's of London, as well as through its captive insurance company, Jupiter Insurance Ltd, which set loss reserves at its policy limit maximum of $700 million.[138] This case did not involve expropriation or investment risk, but rather environmental risk and liability, however it illustrates the operation of domestic U.S. law when the defendant is a foreign corporation.

CASE 5—EXPROPRIATION OF OIL AND GAS ASSETS IN VENEZUELA AND ARGENTINA

Venezuela

This case illustrates the impact of political and ideological change on the security of investor ownership rights, the operation of the ICSID and the options available to claimants and defendants involved in investment disputes by focusing on the expropriations in Venezuela. This case also includes a short digression on expropriation in Argentina. Venezuela nationalized its oil and gas industry in the 1970s, creating the country's state-owned oil and natural gas company, the PDVSA. During the mid-1990s, however, PDVSA implemented a policy referred to as the *Apertura Petrolera* (oil opening), which was intended to mobilize the capital, technology and managerial capabilities of IOCs to maximize the production of crude oil in Venezuela and simultaneously reduce the government revenue needed to finance oil and gas exploration and development in the country.[139]

The "flagship" projects of the *Apertura* were four large projects dedicated to the production, upgrading (i.e., partial refining) and marketing of extra-heavy crude oil from the Orinoco Oil Belt, a large reservoir with an estimated one trillion barrels of heavy (dense and highly viscous) crude oil. Three of these projects (*Petrozuata, Hamaca and Cerro Negro*) are at the center of the arbitration proceedings that ConocoPhillips and ExxonMobil brought against Venezuela at the ICSID in late 2007 in response to Venezuela's nationalization of ConocoPhillips and ExxonMobil's assets in Venezuela.[140]

Timeline of Events

On December 6, 1998, Hugo Chávez, the presidential candidate of the Fifth Republic Movement/United Socialist Party, ran on an anticorruption and antipoverty platform and was elected president of Venezuela.

On November 22, 2001, President Chávez announced the enactment of forty-nine new economic laws, just before the expiration of a law giving legislative power to the president of Venezuela and the executive branch. These laws included new legislation that increased royalty taxes from 16.6 percent to 30 percent on oil production from new oil ventures and mandated the Venezuelan government own more than 50 percent of all new oil ventures.[141]

On April 11, 2002, the first coup attempt against President Chávez occurred. Between December 2002 and February 2003, a general strike was organized by the political opposition to President Chávez to force a new presidential election. The primary economic consequence of the strike was to shut down the Venezuelan oil industry, that is, state-run PDVSA, which provided most of Venezuela's export revenue.[142] The strike lasted approximately two months. The Chávez government responded by firing 19,000 PDVSA employees and replacing them with employees loyal to the Chávez administration.[143]

On October 11, 2004, President Chávez declared that Venezuela was increasing the royalties to be paid by foreign oil companies operating in the Orinoco Basin from 1 percent of the sales price to 16.6 percent on existing oil ventures. Chávez said that this decision marked "the second and true phase of the nationalization of the country's oil." He said the purpose of the increase was to secure "sovereignty" over the country's energy reserves; and that oil prices had undergone a structural (as opposed to a cyclical) change that justified the increase. Political and economic observers outside the country suggested that the government's growing budget deficit was an important factor in the decision to raise royalty rates.[144]

On April 14, 2005, Venezuelan authorities announced a unilateral revision of thirty-two contracts signed with oil companies between 1992 and 1997. These contracts would become JVs in which the state, through PDVSA, would have at least a 51 percent stake in each JV. The Hydrocarbons Law of 2001 had required *new* JVs to include the state oil company, but now the law was being applied retroactively (see entry for November 22, 2001).

Venezuelan authorities also announced that they would increase income taxes from 34 percent to 50 percent in the oil industry, and increase royalties from 16.6 percent to 30 percent on these contracts. This new tax level was applicable to all companies that drilled, produced, operated or processed oil in Venezuela. The Venezuelan government would therefore receive a minimum of 82.5 percent of net income (revenue less production costs).[145] Furthermore, the government made the tax increase from 34 percent to 50 percent retroactive to 2001 and demanded that sixteen foreign oil companies pay back taxes of U.S. $3 billion. These companies included Total (France), BP (United Kingdom), ENI (Italy), Royal/Dutch Shell (The Netherlands), Harvest Vinccler (U.S.), Chevron (U.S.) and Statoil (Norway).[146] Total, Chevron, Statoil and BP agreed to the restructuring.[147]

On June 26, 2007, ExxonMobil and ConocoPhillips refused to comply with the requirement that they grant PDVSA a 51 percent interest in their properties in the Orinoco Basin.[148] Venezuela responded by claiming a majority stake in four oil projects with an estimated total value of $30 billion, operating in the Orinoco river basin (Petrozuata, Hamaca, Cerro Negro and Corocoro).[149]

On September 6, 2007, ExxonMobil registered its claim with the ICSID in response to Venezuela's expropriation of its assets in Venezuela, seeking compensation of $10 billion.[150] On December 13, 2007, ConocoPhillips registered its claim with the ICSID in response to Venezuela's expropriation of its assets in Venezuela, seeking compensation of U.S. $5.5 to $6 billion.[151]

On May 7, 2009, Venezuela's National Assembly passed a law allowing the Venezuelan government to nationalize the assets of certain domestic and foreign oil service companies. "The Organic Law that Reserves to the State the Assets and Services Related to Primary Hydrocarbons Activities" (the "Law") was published in Official Gazette N° 39,173 on May 7, 2009. The purpose of the Law is to reserve in favor of the State of Venezuela those assets and services related to the performance of the primary activities set forth in the Organic Hydrocarbons Law. The reserved activities shall be directly carried out by PDVSA or the affiliated company designated for such purposes or through JV companies under the control of PDVSA or its affiliates."[152]

On May 8, 2009, a gas injection project owned by Williams Companies, a U.S. company, was confiscated by the Venezuelan government.[153] On June 24, 2010, the Venezuelan government seized eleven oil rigs owned by H&P, a U.S. company. H&P's eleven oil rigs had been shut down for several months because Venezuela was unwilling or unable to pay H&P what it was owed.[154] On March 25, 2011, Williams Companies, Inc. filed an international arbitration suit against Venezuela at the ICSID, seeking compensation for the two natural gas compression facilities seized by the Venezuelan government in 2009.[155]

In September 2011, the Venezuelan government said it was willing to pay a combined sum of $2.5 billion to ExxonMobil and ConocoPhillips in their arbitration cases at the ICSID against Venezuela.[156]

On September 30, 2011, H&P filed a lawsuit against Venezuela in the U.S. district court for the District of Columbia for expropriation of its eleven drilling rigs in June 2010. The lawsuit sought $32 million in back payments for unpaid services and several hundred million dollars for the value of its eleven drilling rigs. Citing increasing hostility toward U.S. companies by the Venezuelan government, the company argued that it could not get a fair trial in the Venezuelan courts, where its request for compensation had not progressed or even been answered.[157]

On January 1, 2012 the Paris-based ICC awarded ExxonMobil $907.6 million for breach of contract, substantially less than the $10.0 billion ExxonMobil was seeking. The ICC reduced the $907.6 million award by $160.6 million for liabilities owed by ExxonMobil to Venezuela, making the net award $747 million. In addition, ExxonMobil had previously petitioned a New York court to seize $305 million from a PDVSA bank account, which was subsequently turned over to ExxonMobil. The actual amount Venezuela owed ExxonMobil therefore, was $442 million (907.6 – 160.6 – 305.0 = 442.0), which is the smallest amount ExxonMobil could have expected to receive, because it represented the book value of its 41.7 percent interest in the Cerro Negro partnership.[158] The arbitration in the ICC was intended to determine damages, if any, related to ExxonMobil's claim that Venezuela and PDVSA had breached the contract when it changed its provisions unilaterally and subsequently expropriated the partnership.[159]

However, the case was not finished because ExxonMobil had sought arbitration in two separate courts: the ICC (just discussed) and the ICSID. This was possible under international law, because in the early 1990s, when ExxonMobil decided to participate in the Cerro Negro project, it signed a contract with PDVSA and Venezuela in which ExxonMobil owned a 41.7 percent stake in Cerro Negro through a subsidiary based in the Netherlands; and Venezuela and the Netherlands had signed a bilateral investment treaty to promote and protect investments in both countries. The ICSID arbitration panel was therefore judging whether this treaty had been violated by the Venezuelan government. The tribunal's judgment would be based on the text of the treaty and any violations that may have occurred.[160] The ICSID panel could award ExxonMobil compensation for violations of this treaty, for example, not being paid for amounts owed by Venezuela before the expropriation took place, Venezuela modifying contracts unilaterally, Venezuela discriminating against foreign investors in relation to local investors and not providing prompt and equitable compensation for the assets expropriated.[161]

In January 2012, President Chávez threatened to ignore any ICSID decision regarding the multibillion claim made by ExxonMobil for its nationalized oil projects and he threatened to withdraw from the arbitration panel. Venezuela's energy minister said an agreement could not be reached with ConocoPhillips over the company's expropriated assets.[162]

On January 24, 2012, the World Bank received "written notice of denunciation of the Convention on the Settlement of Investment Disputes between States and Nationals of Other States" (the ICSID Convention) from the *Republica Bolivariana de Venezuela.* The denunciation was to take effect six months after the receipt of Venezuela's notice, that is, on July 25, 2012.[163]

On February 15, 2012 Venezuela's state oil company, PDVSA, claimed that it had paid ExxonMobil Corp. approximately $255 million in compensation for nationalized assets, which was substantially less than the gross amount of $907.6 million ordered by the ICC or even the net amount of $442 million announced by ExxonMobil. President Chávez claimed that the difference of $187 million (442-187 = 255) reflected the debt owed by ExxonMobil to PDVSA resulting from the repurchase of the Cerro Negro bonds. It was not clear what PDVSA meant by this statement because those bonds were repurchased by PDVSA in a decision in which ExxonMobil did not participate.[164]

On March 22, 2012, the Williams Companies, Inc. and the Venezuelan government reached a settlement in which Venezuela agreed to pay Williams Companies $420 million as compensation for its nationalization of the company's assets in Venezuela in May of 2009.[165]

On September 3, 2013, the World Bank's arbitration panel, the ICSID, ruled that Venezuela illegally expropriated ConocoPhillips' Petrozuata, Hamaca and Corocoro projects.[166]

On September 30, 2013, U.S. district judge Robert Wilkins ruled that H&P International Drilling and its Venezuelan subsidiary had standing to pursue its expropriation claim in U.S, courts against the Bolivarian Republic of Venezuela, its state-sponsored energy company, PDVSA and its subsidiary PDVSA Petroleo.[167]

On October 10, 2014, Venezuela announced that it would pay ExxonMobil more than a billion dollars for the nationalization of its operations in Venezuela in 2007 after the ICSID ordered the Venezuelan government to pay

$1.6 billion (1.2 billion euros) to the company. The award was a compensation for the expropriation of the Cerro Negro project, the *La Ceiba* project and "production and export curtailments" imposed on the Cerro Negro development in 2006 and 2007.

ExxonMobil said the decision supported its view that Venezuela failed to give fair compensation at the time of the expropriation. Venezuelan foreign minister Rafael Ramirez said the compensation ordered by the ICSID in the case was within a "reasonable range" and claimed victory in the court case. Ramirez said the tribunal's award was a victory for Venezuelan sovereignty over "exaggerated" claims. He said, however, that Venezuela would pay the fee, only after deducting a previous payment to ExxonMobil made by PDVSA for $908 million related to the Cerro Negro expropriation.[168]

Digression on Argentina

Between 2001 and 2004, natural gas and electricity transmission companies brought fifteen cases against the Argentine government at the ICSID in response to domestic energy pricing policies and tariffs, but not expropriations. Nestor Kirchner served as Argentina's president from May 25, 2003 to December 10, 2007. President Kirchner was considered by some to be a left-wing president.[169] However, he did not propose left-wing policies, such as socialization of production or the nationalization of public services that had previously been privatized during the presidency of Carlos Menem (July 8, 1989 to December 10, 1999). On December 10, 2007, his wife, Cristina Fernandez de Kirchner, succeeded him and began a second term on December 10, 2011.

On April 16, 2012, the Argentine government announced that it would acquire a controlling interest in Yacimientos Petrolíferos Fiscales (YPF) by nationalizing the Spanish oil company Repsol's 57.4 percent ownership of YPF. This takeover followed more than two months of increasing government pressure on YPF after the cost of fuel imports into Argentina doubled to $9.4 billion in 2011.[170] On December 18, 2012, Repsol registered a claim with the ICSID seeking compensation of $10 billion for its 57.4 percent interest in YPF.[171]

On July 30, 2014, Argentina defaulted on $539 million in interest payments due July 30, 2014.[172] On September 13, 2014, Argentina neared default on $200 million in interest payments due September 30, 2014.

On March 1, 2016, Argentina agreed to a $4.65 billion cash payment to its main remaining creditors and was to present the deal to the Argentine Congress for a vote which would end fourteen years of litigation and pave the way for its return to global credit markets.[173] On April 19, 2016 Argentina sold $16.5 billion in sovereign bonds in its first international bond issue since its record 2002 default.[174]

Case Analysis

The expropriations in Venezuela and Argentina illustrate the factors that can lead to resource nationalization. These explanations can include: (1) a change in the ruling party's economic and political ideology, (2) the residual resentment left by colonialism and the exploitation of this resentment by politicians, (3) national budget deficits and (4) political instability. The discussion below examines each of these as they relate to Venezuela and Argentina.

Political Ideology

The socialist ideology of the Chávez and Maduro governments was a major factor in the expropriation of oil and natural gas assets in Venezuela. This is less true in Argentina where the decision to expropriate was driven more by Argentina's worsening budget deficit and increasing debt.

Colonialism and Political Opportunism

The legacy of colonialism is still evident in the degree of income inequality that exists in Venezuela and Argentina. The resentment this inequality produces among the lower socioeconomic classes in both countries allows politicians to exploit this sentiment by encouraging resource nationalism.

National Budgets

Argentina's budget surplus as a percent of GDP was +.2 percent in 2010, and its budget deficit was –1.7 percent in 2011, –2.5 percent in 2012, –2.6 percent in 2013, –3.8 percent in 2014, and –5.5 percent in 2015.[175] Venezuela's budget surplus as a percent of GDP was +3.1 percent in 2007, and its budget deficit was –1.2 percent in 2008, –5.1 percent in 2009, –4.0 percent in 2010, –5.3 percent in 2011, –8.5 percent in 2012, –11.5 percent in 2013,[176] –13.2 percent in 2014 and –16.5 percent in 2015.[177] The increasing budget deficits in Venezuela and Argentina compelled politicians in both countries to seek additional sources of revenue through nationalization of a wide range of private assets, including oil and gas properties.

Political Instability

In 2007, the year in which the largest number of expropriations of oil and gas assets occurred in Venezuela, the country's political stability rating had declined to 13 (0 is least stable and 100 most stable). In 2012, the year in which Repsol's 57.4 percent interest in YPF was expropriated, Argentina's political stability rating declined to 39. The expropriation of the oil companies operating in Venezuela may have been the ruling party's response to its mounting financial problems and its increasing vulnerability to its political opposition and in Argentina to voters seeking improved economic prospects.

In Venezuela, the primary factors leading to expropriation were the socialist orientation of the Chávez government and the political instability of the country. In Argentina, the primary factors leading to expropriation were the worsening budget deficit and the popular appeal of nationalization among Argentina's voters. Again, there is no evidence that the U.S. government intervened on behalf of ConocoPhillips or ExxonMobil in their disputes with the Venezuelan Government supporting the view that national governments seldom intervene on behalf investors.

Efficacy of Joint Ventures and Project Finance

The ExxonMobil and ConocoPhillips' projects in Venezuela *(Petrozuata, Hamaca and Cerro Negro)* employed JVs and project finance. For example, the Petrozuata project was financed with a combination of bank debt and project bonds ($450 million in bank financing and $1.0 billion in project bond financing).[178] The adoption of a JV structure, the participation of commercial banks and the sale of bonds in the private placement market were intended to provide protection against political risk in general and creeping and outright expropriation, specifically. However, ExxonMobil's and ConocoPhillips' experience in Venezuela demonstrates that a nation's sovereignty and its ability to use its armed forces means that risk mitigation techniques will not be effective if the host government chooses to ignore its legal obligations.

CASE 6—THE TNK-BP JOINT VENTURE, OAO YUKOS AND RUSSIAN POLITICS

This case is unique in its specific details, but representative of the sometimes-complex interaction between a foreign investor (BP), Russian investors (AAR), state-owned oil and gas companies (Gazprom and Rosneft) and a national government (Russia).

Investing in the Russian oil and gas industry has been a difficult experience for many foreign oil and natural gas companies, particularly for BP, which sold its 50 percent interest in its TNK-BP joint venture in March 2013. The TNK-BP joint venture encountered numerous problems including the imposition of a billion-dollar claim for back-taxes by the Russian government; delays in the issuance of environmental permits; its offices searched by Russian state police; and its chief executive officer forced to leave Russia because he might be arrested by the Russian authorities. When TNK-BP was sold, it accounted for approximately 25 percent of BP's global oil production.

On June 26, 2003, the agreement establishing the TNK-BP joint venture was signed by BP's chief executive officer, John Browne and chairman of Alfa Group, Mikhail Fridman. Alfa Group was formed after the Soviet Union was dissolved and Russian state-owned industries were privatized. To manage their interests in TNK-BP, Fridman and his Russian partners (Viktor Vekselberg and Leonard Blavatnik) formed a consortium named Alfa-Access-Renova (AAR). BP and AAR agreed that each would hold a 50 percent interest in TNK-BP.

Between 2003 and 2005, TNK-BP's oil and gas production increased by 24 percent, after BP engineers introduced new drilling technology that reversed a long decline in production at TNK-BP's most important oil field, the Samotlor field in western Siberia. However, the partnership between BP and AAR experienced problems almost from the beginning and the relationship became more contentious over time. BP and AAR repeatedly disagreed about the investment strategy the JV should pursue. AAR wanted TNK-BP to invest in oil and gas exploration outside of Russia, for example, Iraq, Lithuania and Turkey, but BP considered TNK-BP to be its Russian subsidiary and did not want TNK-BP competing with BP's projects in other parts of the world. In addition, AAR owned interests in other businesses, including telecommunications, banking, aluminum, media, entertainment and retailing; and wanted TNK-BP to reinvest some of the cash it produced, in those businesses.[179]

In July 2003, in a series of events unrelated to TNK-BP, Platon Lebedev, the fourth largest shareholder in OAO Yukos (a large privately owned oil company in Russia) was arrested on charges of illegally acquiring a stake in the state-owned fertilizer company, *Apatit,* in 1994. The arrest was followed by investigations into tax returns filed by OAO Yukos.[180] In October 2003, Mikhail Khodorkovsky, chairman and CEO of OAO Yukos was arrested and charged with fraud and tax evasion. His arrest and subsequent conviction on these charges were widely interpreted to mean that the Kremlin intended to reign in the so-called oligarchs, who had become rich during the period of privatization in Russia in the 1990s and were becoming a political threat.[181]

In October 2005, Russian tax authorities imposed a claim for $936 million in back taxes on TNK-BP, which temporarily raised the prospect of expropriation by the Russian government. However, the possibility of expropriation diminished after BP's CEO, John Browne, met with Russian president, Vladimir Putin, and President Putin praised BP as a "good corporate citizen." The claims for back taxes were later substantially reduced.[182]

On October 3, 2006, in another series of events unrelated to TNK-BP, other private sector oil companies that had invested in Russia also came under pressure from the Russian government. For example, after being accused of environmental violations and threatened with multibillion-dollar penalties, Shell Sakhalin Holdings B.V. (a subsidiary of Royal Dutch Shell), Mitsui Sakhalin Holdings B.V. (a subsidiary of Mitsui) and Diamond Gas Sakhalin (a subsidiary of Mitsubishi), agreed to give up controlling interest in the

Sakhalin-2 offshore oil and gas development project. *Gazprom*, the state-owned gas producer, acquired majority ownership in Sakhalin-2; and Royal Dutch Shell and its partners Mitsui and Mitsubishi had their ownership in the project reduced. These developments made BP's continuing 50 percent ownership in TNK-BP an exception in an otherwise rapidly renationalizing Russian oil and gas industry.[183]

On March 19, 2008, Officers from Russia's Interior Ministry raided BP's and TNK-BP's offices in Moscow related to several new investigations for tax evasion. The investigations had begun in 2007 and gained momentum in early 2008. By April 2008, the ministry was again investigating claims for back taxes of more than $900 million against various TNK-BP subsidiaries. In addition, the Federal Security Service (FSB) arrested an employee of TNK-BP and his brother, an independent energy consultant, and charged them with industrial espionage.[184]

In April 2008, Tetlis, (a Moscow based brokerage firm and one of a small group of minority shareholders in TNK-BP Holdings), filed a lawsuit against TNK-BP. Tetlis sued TNK-BP in a Siberian court over a longstanding agreement that allowed technical specialists from BP to be "seconded" to TNK-BP. (When an employee is "seconded," the person remains an employee of their current employer (BP), but a contractual agreement is entered between the employing organization and a third-party company [TNK-BP] under which the employee will perform work for the benefit of the third-party company [TNK-BP]. The "seconded" employees are never employed by the third-party company.) The Siberian court issued an injunction that prevented 148 "seconded" BP employees from entering the offices of TNK-BP.

BP and AAR also became engaged in a public argument regarding the salaries of BP employees working at TNK-BP. AAR insisted that the BP's expatriate employees were unnecessary and overpaid, but BP asserted that they were necessary and were not overpaid. Furthermore, TNK-BP's chief executive officer, Bob Dudley accused AAR of interfering with the Russian work permit application process to prevent BP's employees from renewing their Russian visas, forcing them to leave the country when their old visas expired.[185]

In May 2008, the FSB conducted another raid on BP's Moscow office and in a separate action Russia's Natural Resources Ministry announced an environmental inspection of TNK-BP's Samotlor oil field. In July 2008, in a separate lawsuit, a group of Russian managers working at TNK-BP accused TNK-BP chief executive officer, Bob Dudley, of discrimination, claiming that TNK-BP's Western employees were paid much more than local Russian staff. Eventually, the 148 "seconded" employees were forced to leave Russia and dozens of other Western employees not directly employed by TNK-BP had to leave the country because they were unable to renew their visas. One of them was TNK-BP's chief executive officer, Bob Dudley, who left on July 24, 2008, ending a long struggle with AAR and the Russian authorities.[186]

In September 2008, BP reached an agreement with AAR, in which it ceded operating control of the TNK-BP JV to the Russian shareholders (AAR). Under the terms of the agreement finalized four months later, Dudley resigned as TNK-BP's chief executive officer and AAR obtained the right to approve his replacement. The two sides also agreed to increase the size of TNK-BP's board of directors and add three independent directors, fulfilling yet another key demand of AAR.[187]

TNK-BP's legal problems significantly diminished after the agreement with AAR was reached. In September 2008, the Russian Interior Ministry announced that TNK-BP had settled most of the claims against it for back taxes, and the ministry eventually ended its investigation for tax evasion against TNK-BP. In addition, the two brothers arrested by the FSB for industrial espionage were given suspended sentences and Tetlis and the Russian managers dropped their discrimination lawsuits.[188]

In 2009, BP played a more passive role in TNK-BP, allowing AAR to run the JV. However, that changed on April 20, 2010, when an explosion on BP's Deepwater Horizon drilling rig in the Gulf of Mexico produced a massive oil spill that resulted in billions of dollars in fines by the U.S. government, cleanup costs and liability to individuals, businesses and states along the Gulf Coast. On July 27, 2010, three months after the explosion and oil spill, BP announced that Bob Dudley would succeed Tony Hayward as BP's chief executive officer on October 1, 2010. Dudley was also appointed to the board of directors of BP.[189]

In January 2011, Dudley announced a strategy that was intended to stabilize BP's financial position and give the company access to significant undeveloped oil and gas reserves in Russia. The plan proposed an exchange of common shares between BP and *Rosneft* (a Russian state-owned oil company) in which *Rosneft* would take a 5 percent stake in BP in exchange for BP taking a 9.5 percent stake in Rosneft. In addition, BP and Rosneft agreed to explore and develop three license blocks in the Kara Sea, a Russian part of the Arctic Ocean. However, the proposal came to nothing because AAR, which was upset that BP had not thought to include them in the Arctic deal, cited the original BP-AAR shareholder agreement, which required that BP pursue any Russian projects exclusively through TNK-BP. AAR sued to block the BP-Rosneft deal and a London court and a Stockholm arbitration tribunal agreed that BP was violating the shareholder agreement. Despite last-minute negotiations with AAR to preserve the deal with *Rosneft,* discussion of the proposed alliance ended in May 2011.[190]

In October 2012, BP and AAR simultaneously announced they would sell their respective 50 percent interests in TNK-BP to *Rosneft,* a deal, that valued the JV at $55 billion. On March 21, 2013, BP and AAR sold their respective 50 percent interests in TNK-BP to the Russian state-owned oil company *Rosneft*, in a deal that closed on March 21, 2013. BP received compensation of $26.7 billion ($12.5 billion in cash and 19.75 percent of Rosneft common stock valued at $14.2 billion). AAR received $27.7 billion in cash from Rosneft.[191, 192]

Case Analysis

The role of the Russian government in the disputes between BP and AAR is unclear. Some observers do not believe that the Russian government led the legal campaign against TNK-BP; rather, they believe that AAR used its connections in Russia's government ministries and agencies to apply pressure on Dudley and his team, whom they disliked.[193]

> The security services were used by AAR as one of the means of achieving their main goal. "They (AAR) wanted to seize operational control of the company and squeeze out Dudley, who was a very major irritant for them. . . . It's no secret that they had the ability to do this, since they had tight relationships with very senior individuals in the Russian leadership."[194]
>
> BP was not properly prepared for the pressure. It reacted slowly, sometimes even passively, since it is a huge and risk-averse bureaucratic machine that functions relatively well in normal times, but at the time it was not fully up to the task of outright corporate warfare in the Russian style.[195]

The Russian oil and natural gas industry had been renationalizing under President Putin since his election. The evidence for this includes: (1) The intentional bankruptcy and subsequent sale of Yukos's assets to the state-owned company *Rosneft* (2) the allegation of threats to marine life and other environmental problems at Royal Dutch Shell's Sakhalin-2 site and (3) a subsequent injunction that rescinded permits for Sakhalin-2's second phase and paved the way for *Gazprom's* eventual acquisition of a controlling interest in the Sakhalin-2 project. The Russian ministries and Putin himself, therefore, did not object when they were asked to put pressure on TNK-BP.

In the case of OAO Yukos the charges against Mikhail Khodorkovsky and Platon Lebedev and the intentional bankruptcy of the company were motivated primarily by Vladimir Putin's desire to remove two businessmen who were critical of his regime and therefore a threat to his political power and re-election. In the case of TNK-BP neither BP nor AAR presented a threat to Putin's continuance in office, but TNK-BP was a large and successful project that had significant economic and geopolitical value to the Russian government. There is no evidence in the BP-TNK and Royal Dutch Shell cases that the British government or the Dutch government intervened on behalf of BP or Shell in their disputes with Russian regulatory authorities.

NOTES

1. A Barrel Full, Frade Oil Field, Summary Information, http://abarrelfull.wikidot.com/frade-oil-field. Last accessed on August 20, 2017.

2. Andrew Inkpen and Michael H. Moffett, *The Global Oil and Gas Industry—Management, Strategy and Finance.* (Tulsa, OK: PennWell Corporation, 2011) p. 129.

3. Ibid, p. 131.

4. SUBSEAIQ Offshore Oil Field Development, http://www.subseaiq.com/data/Project.aspx?project_id=313&AspxAutoDetectCookieSupport=1, Last updated September 13, 2013.

5. Ibid.

6. Gulf Oil and Gas, "Oil Spill of Frade Field in Brazil—Special Report," http://www.gulfoilandgas.com/webpro1/projects/3dreport.asp?id=104359, Last accessed on May 7, 2014.

7. Ibid.

8. Ibid.

9. Ibid.

10. Ibid.

11. Ibid.

12. SUBSEAIQ Offshore Oil Field Development, http://www.subseaiq.com/data/Project.aspx?project_id=313&AspxAutoDetectCookieSupport=1.

13. Ibid.

14. Gulf Oil and Gas, "Oil Spill of Frade Field in Brazil—Special Report," http://www.gulfoilandgas.com/webpro1/projects/3dreport.asp?id=104359.

15. Ibid.

16. SUBSEAIQ Offshore Oil Field Development, http://www.subseaiq.com/data/Project.aspx?project_id=313&AspxAutoDetectCookieSupport=1, Last updated September 13, 2013.

17. Gulf Oil and Gas, "Oil Spill of Frade Field in Brazil—Special Report," http://www.gulfoilandgas.com/webpro1/projects/3dreport.asp?id=104359.

18. SUBSEAIQ Offshore Oil Field Development, http://www.subseaiq.com/data/Project.aspx?project_id=313&AspxAutoDetectCookieSupport=1.

19. Gulf Oil and Gas, "Oil Spill of Frade Field in Brazil—Special Report," http://www.gulfoilandgas.com/webpro1/projects/3dreport.asp?id=104359.

20. Ibid.

21. Ibid.

22. Ibid.

23. Ibid.

24. Ibid.

25. Upstream, "Chevron OK'd for Full Frade Restart," April 1, 2014, http://www.upstreamonline.com/epaper/article1357081.ece.

26. SUBSEAIQ Offshore Oil Field Development, http://www.subseaiq.com/data/Project.aspx?project_id=313&AspxAutoDetectCookieSupport=1.

27. Exxon Valdez Oil Spill Trustee Council. "Questions and Answers," http://www.evostc.state.ak.us/index.cfm?FA=facts.QA, http://www.evostc.state.ak.us/index.cfm?FA=facts.QA.

28. U. S. Coast Guard. "*On Scene Coordinator Report on Deepwater Horizon Oil Spill*," September 2011. http://www.uscg.mil/foia/docs/dwh/fosc_dwh_report.pdf.

29. Andrew Inkpen and Michael H. Moffett, *The Global Oil and Gas Industry—Management, Strategy and Finance.* (Tulsa, OK: PennWell Corporation, 2011) p. 351.

30. Chevron, https://www.chevronaustralia.com/our-businesses/gorgon.

31. Andrew Inkpen and Michael H. Moffett, *The Global Oil and Gas Industry – Management, Strategy and Finance.* (Tulsa, OK: PennWell Corporation, 2011), p. 352.

32. SUBSEAIQ Offshore Field Development, http://www.subseaiq.com/data/Project.aspx?project_id=271.

33. Ibid.

34. Ibid.

35. Ibid.

36. Ibid.

37. Ibid.

38. Ibid.

39. Ibid.

40. Ibid.

41. Ibid.

42. Ibid.

43. Ibid.

44. Ian Wilson, "Chevron's Gorgon Carbon Project on Track for Injection in 2015," *Bloomberg BusinessWeek,* October 23, 2012, http://www.businessweek.com/news/2012-10-23/chevron-s-gorgon-carbon-project-on-track-for-injection-in-2015.

45. Matt Chambers, "Chevron ups Cost Estimate on Gorgon LNG Project to $52 Billion," *The Australian*, December 6, 2012, http://www.theaustralian.com.au/business/mining-energy/chevron-ups-cost-estimate-on-gorgon-lng-project-to-52bn/story-e6frg9df-1226530997643.

46. Kathryn Diss and Sue Lannin, "Chevron Reveals Second Gorgon Gas Project Cost Blowout," December 12, 2013, http://www.abc.net.au/news/2013-12-12/gorgon-gas-project-cost-blowout/5151982.

47. Ibid.

48. Maritime Union of Australia, "Global Campaign to Target Chevron over Gorgon Delays and Overruns," April 9, 2014, http://www.mua.org.au/news-story/3852/.

49. The Australian, "Chevron's Gorgon Project 78 percent Complete," March 12, 2014, http://www.theaustralian.com.au/business/latest/chevron-eyes-demand-growth/story-e6frg90f-1226852302166?nk=44368b41aa00c57e0356ec94a1c64a47.

50. Chevron, "Gorgon Project Update," August 1, 2014, http://www.chevron.com/media/gorgon/Gorgon-Progress-Update-08012014.pdf.

51. Reuters, "Struggling to Lock in Buyers, Chevron's Gorgon Casts doubt on Global LNG Sales Model," August 25, 2014, http://www.reuters.com/article/2014/08/25/chevron-lng-idUSL4N0QQ22X20140825.

52. Anjli Raval, "Brent Crude Falls Under $50 per Barrel," *Financial Times*, January 7, 2015, http://www.ft.com/cms/s/0/d04d5e86-9644-11e4-a40b-00144feabdc0.html#axzz3O9da8H3P.

53. Andrew Inkpen and Michael H. Moffett, *The Global Oil and Gas Industry – Management, Strategy and Finance*. (Tulsa, OK: PennWell Corporation, 2011), p. 411.

54. Hydrocarbons technology.com Baku-Tbilisi-Ceyhan (BTC) pipeline. Last accessed August 20, 2017, http://www.hydrocarbons-technology.com/projects/bp/

55. Trend, "Tankers Carrying Oil Delivered by BTC Pipeline Shipped from Turkey," April 21, 2014, http://en.trend.az/capital/energy/2265951.html.

56. *Turkish Daily News*, "Timeline of the Baku-Tbilisi-Ceyhan Pipeline," July 13, 2006, http://www.hurriyetdailynews.com/default.aspx?pageid=438&n=timeline-of-the-baku-tbilisi-ceyhan-pipeline-2006-07-13.

57. Zeyno Baran, "The Baku-Tbilisi-Ceyhan Pipeline: Implications for Turkey," (PDF), 103–118, http://www.silkroadstudies.org/BTC_6.pd.f S. Frederick Starr and Svante E. Cornell, "The Baku-Tbilisi-Ceyhan Pipeline: Oil Window to the West," *The Central Asia-Caucasus Institute*, Silk Road Studies Program, 2005, http://www.silkroadstudies.org/BTC.pdf.

58. Ibid.

59. Ibid.

60. Corner House, "The Legal Regime for the Baku-Tbilisi-Ceyhan (BTC) Oil Pipeline Project," March 11, 2011, http://www.thecornerhouse.org.uk/sites/thecornerhouse.org.uk/files/Implications percent20of percent20NCP percent20Final percent20Statement percent20BP_0.pdf.

61. Ibid.

62. Ibid.

63. Alexander's Gas & Oil Connections, "Baku-Tbilisi-Ceyhan Pipeline Company Founded," August 30, 2002, http://www.gasandoil.com/news/central_asia/06e64ce97241a092b52d5a35d6be6a60.

64. International Finance Corporation, *The Baku-Tbilisi-Ceyhan (BTC) Pipeline Project,* Lessons of Experience, September 2006, No. 2, IFC, World Bank Group, http://www.gcgf.org/wps/wcm/connect/d01d2180488556f0bb0cfb6a6515bb18/BTC_LOE_Final.pdf?MOD=AJPERES&CACHEID=d01d2180488556f0bb0cfb6a6515bb18.

65. BP, "BTC Celebrates Full Commissioning," Press release, July 13, 2006, http://www.bp.com/en/global/corporate/press/press-releases/btc-celebrates-full-commissioning.html.

66. BP, "Baku-Tbilisi-Ceyhan Pipeline," http://www.bp.com/en_az/caspian/operationsprojects/pipelines/BTC.html.

67. Trend, "Tankers Carrying Oil Delivered by BTC Pipeline Shipped from Turkey," April 21, 2014, http://en.trend.az/capital/energy/2265951.html.

68. BP, "Baku-Tbilisi-Ceyhan Pipeline," http://www.bp.com/en_az/caspian/operationsprojects/pipelines/BTC.html.

69. International Finance Corporation, *The Baku-Tbilisi-Ceyhan (BTC) Pipeline Project,* Lessons of Experience, September 2006, No. 2, IFC, World Bank Group, http://www.gcgf.org/wps/wcm/connect/d01d2180488556f0bb0cfb6a6515bb18/BTC_LOE_Final.pdf?MOD=AJPERES&CACHEID=d01d2180488556f0bb0cfb6a6515bb18.

70. Transocean, "Transocean's Ultra-Deepwater Semisubmersible Rig Deepwater Horizon Drills World's Deepest Oil and Gas Well," Transocean (September

2, 2009), http://globenewswire.com/news-release/2009/09/02/403900/172461/en/Photo-Release-Transocean-s-Ultra-Deepwater-Semisubmersible-Rig-Deepwater-Horizon-Drills-World-s-Deepest-Oil-and-Gas-Well.html.

71. Guy Crittenden, "Understanding the Initial Deepwater Horizon Fire," *HazMat Management,* May 10, 2010, http://www.hazmatmag.com/news/understanding-the-initial-deepwater-horizon-fire/1000370689/?&er=NA; USA Today, "Gulf Oil Spill Now Largest Offshore Spill in History as BP Continues Plug Effort," May 27, 2010, http://www.cacoastkeeper.org/news/gulf-oil-spill-now-largest-in-us-history-as-bp-continues-plug-effort.

72. U.S. Coast Guard, "On Scene Coordinator Report on Deepwater Horizon Oil Spill," September 2011, http://www.uscg.mil/foia/docs/dwh/fosc_dwh_report.pdf.

73. Jennifer Larino, "Federal Judge Rejects BP's bid to lower $13.7 billion Oil Spill Fine," NOLA.com; Times-Picayune, February 19, 2015, http://www.nola.com/business/index.ssf/2015/02/bp_loses_bid_to_cap_oil_spill.html.

74. William Welch and Chris Joyner, "Memorial Services Honors 11 Dead Oil Rig Workers," *USA Today,* May 25, 2010, http://usatoday30.usatoday.com/news/nation/2010-05-25-oil-spill-victims-memorial_N.htm.

75. CNN, "Oil Slick Spreads from Sunken Rig," *(video interview),* April 22, 2010, http://www.cnn.com/2010/US/04/22/oil.rig.explosion/.

76. CBC News, "Oil Rig Wreck Leaks into Gulf of Mexico," Associated Press, April 25, 2010, http://www.cbc.ca/news/world/oil-rig-wreck-leaks-into-gulf-of-mexico-1.911697.

77. David Barstow, David Rohde and Stephanie Saul, "Deepwater Horizon Rig Disaster," *The Wall Street Journal*, December 25, 2010, http://www.nytimes.com/2010/12/26/us/26spill.html?pagewanted=all.

78. Associated Press, Coast Guard New Oil Leak from Area Where Rig Exploded, Sank in Gulf; Spill Heads to Coast (online video), April 28, 2010, http://www.youtube.com/watch?v=WhUjqi7IlzY.

79. BBC News, "US Military Joins Gulf of Mexico Oil Spill Effort," April 29, 2010, http://news.bbc.co.uk/2/hi/americas/8651624.stm.

80. Tim Webb, "BP Boss Admits Job on the Line over Gulf Oil Spill," *The Guardian,* May 13, 2010, http://www.theguardian.com/business/2010/may/13/bp-boss-admits-mistakes-gulf-oil-spill.

81. Market Wired, "Transocean Ltd. Affiliates File Limitation of Liability Petition," (Press release), Transocean (May 13, 2010), http://www.marketwired.com/press-release/transocean-ltd-affiliates-file-limitation-of-liability-petition-nyse-rig-1259747.htm.

82. Wall Street Journal, "Deepwater Horizon Rig Disaster," May 19, 2010, http://online.wsj.com/news/articles/SB10001424052748704302304575213883555525958.

83. Loren Steffy, "New Estimates on Offshore Drilling Ban Costs," *Houston Chronicle*, June 2, 2010, http://blog.chron.com/lorensteffy/2010/06/new-estimates-on-offshore-drilling-ban-costs/.

84. CBS News, "Obama Extends Moratorium on Offshore Drilling," May 27, 2010, http://www.cbsnews.com/news/obama-extends-moratorium-on-offshore-drilling/.

85. Bloomberg News, "BP, Transocean Lawsuits Surge as Oil Spill Spreads in Gulf," May 1, 2010, http://www.bloomberg.com/apps/news?pid=newsarchive&sid=afgkFlcmKJwU.

86. National Park Service, "Response to Oil on Gulf Island Beaches Continues," (Press release), June 4, 2010, http://www.nps.gov/aboutus/oil-spill-response.htm.

87. Jane Ross, "Florida Coast Suffers First Impact from Oil Spill," *Reuters,* June 4, 2010, http://www.reuters.com/article/2010/06/04/us-oil-spill-florida-idUSTRE6533S820100604.

88. Jessica Resnick-Ault, "BP Oil-Leak Estimate Doubled by U.S. Science Panel (Correct)," *BusinessWeek*, June 11, 2010, http://www.bloomberg.com/apps/news?pid=newsarchive&sid=a0uIOSekfLIo.

89. Jonathan Weisman and Guy Chazan, "BP Halts Dividend, Agrees to $20 Billion Fund for Victims," *The Wall Street Journal*, June 17, 2010, http://online.wsj.com/news/articles/SB10001424052748704198004575310571698602094.

90. Michael Kunzelman, "Oil Spewing Once Again in the Gulf," *Myrtle Beach News Online Associated Press,* June 24, 2010, http://www.myrtlebeachonline.com/2010/06/24/1550150/oil-spewing-once-again-in-the.html.

91. Environmental Law Institute, *Deepwater Horizon Oil Spill Litigation Database*, http://www.eli.org/deepwater-horizon.

92. MDL No. 2179 Oil Spill by the Oil Rig "DeepWater Horizon," August 10, 2010, http://www.laed.uscourts.gov/OilSpill/OilSpill.htm.

93. Los Angeles Times, "BP Blown-Out Well Finally Killed at Bottom of Gulf of Mexico," September 19, 2010, http://latimesblogs.latimes.com/greenspace/2010/09/blown-out-bp-well-finally-killed-at-bottom-of-gulf-of-mexico.html.

94. U.S. Department of Justice United States of America v. BP Exploration & Production Inc. et al., Civ. Action No. 2:10-cv-04536, *United States District Court—Eastern District of Louisiana,* http://www.law.uh.edu/faculty/thester/courses/Environmental-Practicum-2014/BP_MOEX_CD_FINAL.pdf.

95. John Wyeth Griggs, "BP Gulf of Mexico Spill Energy," *Law Journal*, Vol. 32, 57–79, 2011, http://felj.org/sites/default/files/docs/elj321/14_57_bp_gulf_of_mexico.pdf

96. Leo King, "BP £24 billion Lawsuits Claim Contractors Failed to use Modelling Software Properly," *Computerworld*, UK (April 21, 2011), http://www.computerworlduk.com/news/it-business/3275978/bp-24bn-lawsuits-claim-contractors-failed-to-use-modelling-software-properly/.

97. Laurel Brubaker Calkins and Jeff Feeley, *Insurance Journal*, (March 3, 2014), http://www.insurancejournal.com/news/national/2014/03/03/322103.htm.

98. United States District Court—Eastern District of Louisiana, http://www.stuarthsmith.com/wp-content/uploads/2012/09/7229-2.pdf.

99. David Ingram, "Deepwater Horizon: U.S. Ramps up Rhetoric on BP over Oil Spill," *Reuters,* September 4, 2012, http://www.theguardian.com/business/2012/sep/05/deepwater-horizon-us-bp-oil-spill.

100. Sam Stein, "BP's Influence Peddling Bears Fruit in Congress," *Huffington Post*, March 12, 2012, http://www.theguardian.com/business/2012/sep/05/deepwater-horizon-us-bp-oil-spill.

101. Sam Stein, "BP's Influence Peddling Bears Fruit In Congress," *Huffington Post,* 12 March 2012, http://www.huffingtonpost.com/2012/03/12/bp-oil-spill-gulf-of-mexico-oil-lobbyists_n_1335556.html.

102. CNN, "DOJ Accuses BP of 'Gross Negligence' in Gulf Oil Spill," September 5, 2012, http://money.cnn.com/2012/09/05/news/companies/bp-oil-spill/.

103. Mira Oberman, "BP Vows to 'vigorously defend' itself at US Oil Spill Trial," *Agence France-Press,* 19 February 2013, http://www.globalpost.com/dispatch/news/afp/130219/bp-vows-vigorously-defend-itself-at-us-oil-spill-trial.

104. Richard Thompson, "BP Oil Spill Trial Continues as Demonstrators Note Upcoming 3-year Anniversary of Disaster," *NOLA.com and The Times-Picayune,* April 16, 2013, http://www.nola.com/news/gulf-oil-spill/index.ssf/2013/04/bp_oil_spill_draws_demonstrato.html.

105. Paul Waldie, "BP turned to Wikipedia to Estimate Size of Spill, U.S. Alleges," *The Globe and Mail* (Toronto), November 15, 2012, http://archive.today/ESugm.

106. Clifford Krauss and John Schwartz, "BP Will Plead Guilty and Pay Over $4 Billion," *The New York Times,* http://www.nytimes.com/2012/11/16/business/global/16iht-bp16.html?pagewanted=all&_r=0.

107. Suzanne Goldenberg and Dominic Rushe, "BP to pay $4.5bn Penalty over Deepwater Horizon Disaster," *The Guardian* (London), November 15, 2012, http://www.theguardian.com/environment/2012/nov/15/bp-deepwater-horizon-gulf-oil-spill.

108. Eyder Peralta, "Government Files First Criminal Charges in BP Oil Spill," *NPR,* April 24, 2012, http://www.npr.org/blogs/thetwo-way/2012/04/24/151284311/first-criminals-charges-in-bp-oil-spill-filed-against-a-bp-engineer.

109. Suzanne Goldenberg and Dominic Rushe, "BP to pay $4.5bn Penalty over Deepwater Horizon Disaster," *The Guardian* (London), November 15, 2012, http://www.theguardian.com/environment/2012/nov/15/bp-deepwater-horizon-gulf-oil-spill.

110. Paul Waldie, "BP turned to Wikipedia to Estimate Size of Spill, U.S. Alleges," *The Globe and Mail* (Toronto), November 15, 2012, http://archive.today/ESugm.

111. Eyder Peralta, "Government Files First Criminal Charges in BP Oil Spill," *NPR,* April 24, 2012, http://www.npr.org/blogs/thetwo-way/2012/04/24/151284311/first-criminals-charges-in-bp-oil-spill-filed-against-a-bp-engineer.

112. John Rudolf, "Kurt Mix, BP Engineer, Faces First Oil Spill Charges," *Huffington Post,* 24 April 2012, http://www.huffingtonpost.com/2012/04/24/kurt-mix-bp-engineer-oil-spill_n_1449316.html.

113. U.S. Environmental Protection Agency, "Transocean Agrees to Plead Guilty to Environmental Crime and Enter Civil Settlement to Resolve U.S. Clean Water Act Penalty Claims from Deepwater Horizon Incident," January 3, 2013, http://yosemite.epa.gov/opa/admpress.nsf/2467feca60368729852573590040443d/ea4c5daf4e864d6985257ae80062d9ab!OpenDocument.

114. Richard Thompson, "Federal judge approves BP Gulf Oil Spill Medical Settlement," *The Times-Picayune,* January 11, 2013, http://www.nola.com/business/index.ssf/2013/01/federal_judge_approves_bp_gulf.html.

115. Katherine Schmidt, "Macondo Trial Gets Under Way," *Upstream Online* (NHST Media Group), February 25, 2013, http://www.upstreamonline.com/live/article1318253.ece.

116. Jef Feeley, Jef and Allen Johnson, Jr., "BP, Transocean Accused of 'Reckless' Actions in Spill," *Bloomberg*, Allen (February 26, 2013), http://www.bloomberg.com/news/2013-02-25/bp-oil-spill-errors-trial-begins-in-new-orleans-court.html.

117. Murrey Jacobson, "By the Numbers: The Oil Spill and BP's Legal Troubles," *PBS News Hour, the Rundown*, July 9, 2013, http://www.pbs.org/newshour/rundown/gulf-oil-spill-by-the-numbers/.

118. Clifford Krauss, "In BP Trial, the Amount of Oil Lost Is at Issue," *New York Times,* September 29, 2013, http://www.nytimes.com/2013/09/30/business/energy-environment/bp-trial-in-2nd-phase-to-set-amount-of-oil-spilled.html?pagewanted=all.

119. Michael Kunzelman, "Claims Against BP Contractors Dismissed at Trial," *The Associated Press,* 20 March 2013, http://bigstory.ap.org/article/plaintiffs-rest-trial-over-2010-gulf-oil-spill, Retrieved 13 April 2013.

120. Deepwater Horizon Claims Center, Claims Administrator Resumes Processing of All BEL Claims, March 1, 2013. http://www.deepwaterhorizoneconomicsettlement.com/docs/BROWNGREER-421750-v3-Alert-Resume_Processing_All_BEL_Claims.pdf.

121. Insurance Business, "BP: Insurance Policy Should Cover us for Gulf Oil Spill," November 25, 2013, http://www.ibamag.com/news/bp-insurance-policy-should-cover-us-for-gulf-oil-spill-16271.aspx.

122. John Schwartz, "BP Accuses Texas Lawyer of 'Brazen Fraud' in Workers' Claims Over Gulf Oil Spill," *The New York Times*, December 17, 2013, http://www.nytimes.com/2013/12/18/us/bp-accuses-texas-lawyer-of-brazen-fraud-in-workers-claims-over-gulf-oil-spill.html?_r=0.

123. The Economist, "A Shrunken Giant," February 8, 2014, http://www.economist.com/news/business/21595935-british-oil-company-safer-smaller-sadder-and-wiser-its-disaster-gulf.

124. Laurel Calkins and Jef Feeley, "BP Must Live with $9.2 Billion Oil-Spill Deal, Court Says," *Bloomberg News*, March 4, 2014, http://www.bloomberg.com/news/2014-03-04/bp-must-live-with-oil-spill-settlement-terms-court-rules-1-.html.

125. Ed Crooks, "BP Refuses to Fund Gulf Oil Spill Studies," *Reuters,* April 20, 2014, http://www.ft.com/intl/cms/s/0/cdf8ba8c-c8a3-11e3-8976-00144feabdc0.html#axzz31WtwsjmY.

126. The Times-Picayune, "Appeals Court will not re-hear BP Settlement Issue," *Associated Press,* May 19, 2014, http://www.nola.com/environment/index.ssf/2014/05/appeals_court_wont_rehear_bp_s.html.

127. Kathryn Sayre, "U.S. Refuses for now, to Stop, BP Oil Spill Payments," *The Times Picayune,* June 9, 2014, http://www.nola.com/business/index.ssf/2014/06/supreme_court_refuses_to_stop.html.

128. Nick Snow, Halliburton Announces $1.1 Billion Macondo Claims Agreement, *Oil & Gas Journal,* September 3, 2014, http://www.ogj.com/articles/2014/09/halliburton-announces-1-1-billion-macondo-claims-agreement.html.

129. The New Orleans Sun, "BP Found 'grossly negligent' in Gulf of Mexico Oil Spill," September 2, 2014, http://www.neworleanssun.com/index.php/sid/225407625.

130. Forbes, "BP 'Grossly Negligent' In Gulf Spill, Eyes $18 Billion Penalty–And Tax Deduction," September 4, 2014, http://www.forbes.com/sites/

robertwood/2014/09/05/bp-grossly-negligent-in-gulf-spill-eyes-18-billion-penalty-and-tax-deduction/.

131. Katherine Rushton and Emily Gosden, "Reckless Oil Giant could Face Up to $18bn in Clean Water Act fines for 2010 Disaster after Judge's Ruling, which Sends Shares Down 6pc," *The Telegraph*, September 4, 2014, http://www.telegraph.co.uk/finance/newsbysector/energy/oilandgas/11075316/BP-faces-up-to-18bn-fines-after-Gulf-of-Mexico-gross-negligence-ruling.htmlBP faces up to $18bn fines after Gulf of Mexico 'gross negligence' ruling.

132. Katherine Rushton, "BP Calls for Trial in Deep Water Horizon Case," The Telegraph, October 3, 2014, http://www.telegraph.co.uk/finance/newsbysector/energy/oilandgas/11139068/BP-calls-for-retrial-of-Deepwater-Horizon-case.html.

133. Reuters, "US Supreme Court Rejects BP Challenge to Gulf Spill Settlement," *The Telegraph*, December 8, 2014, http://www.telegraph.co.uk/finance/newsbysector/energy/oilandgas/11280306/US-Supreme-Court-rejects-BP-challenge-to-Gulf-spill-settlement.html.

134. IRMI, February 13, 2015, https://www.irmi.com/articles/expert-commentary/in-re-deepwater-horizon-additional-insured-questions-resolved.

135. Jennifer Larino, "Federal Judge Rejects BP's bid to lower $13.7 billion Oil Spill Fine," NOLA.com; *Times-Picayune*, February 19, 2015, http://www.nola.com/business/index.ssf/2015/02/bp_loses_bid_to_cap_oil_spill.html.

136. U.S. Department of Justice, Press Release, "U.S. and Five Gulf States Reach Historic Settlement with BP to Resolve Civil Lawsuit over DeepWater Horizon Oil Spill," October 5, 2016.

137. Oil & Gas Journal, "BP's Total Liability for Deepwater Horizon Surpasses $60 Billion," July 15, 2016, http://www.ogj.com/articles/2016/07/bp-s-total-liability-for-deepwater-horizon-surpasses-60-billion.html.

138. Derek A. Jones and Jason B. Kurtz, "Deepwater Horizon: Torrent of Oil, Flood of Insurance Issues," July 23, 2010, http://www.milliman.com/insight/pc/Deepwater-Horizon-Torrent-of-oil–flood-of-insurance-issues/.

139. Juan Carlos Boue, "ConocoPhillips and ExxonMobil v. Venezuela: Using Investment Arbitration to Rewrite a Contract," *International Institute for Sustainable Development,* September 20, 2013, http://www.iisd.org/itn/2013/09/20/conoco-phillips-and-exxon-mobil-v-venezuela-using-investment-arbitration-to-rewrite-a-contract/.

140. Ibid.

141. The Economist, "To the Barricades," (November 22, 2001); Financial Times, "Venezuelan Oil Law Threatens Investment," (November 20, 2001); Business News Americas, "Silva Holds firm on Hydrocarbons Law," May 27, 2002, http://www.bnamericas.com/news/oilandgas/Silva_holds_firm_on_Hydrocarbons_Law.

142. Bart Jones, Bart, *Hugo! The Hugo Chavez Story from Mud Hut to Perpetual Revolution.* (London: The Bodley Head, 2008).

143. UN News Centre, "Repression in Myanmar, China, Colombia, Venezuela," March 30, 2004, http://www.scoop.co.nz/stories/WO0403/S00397.htm.

144. BBC News, "Venezuela Raises Oil Drilling Tax," October 11, 2004, http://news.bbc.co.uk/2/hi/americas/3732224.stm.

145. Financial Post – Canada, "Venezuela Announces Plan to Increase Income Tax on Private Oil Projects," April 18, 2005.

146. Business News Americas, "Seniat to Close Offices of 2 Foreign Oilfield Operators," 4 November 2005, http://www.bnamericas.com/news/oilandgas/Seniat_to_close_offices_of_2_foreign_oilfield_operators.

147. Business News Americas, "Petrobras Reaffirms Investments Despite Contract Changes," 22 November 2005, http://www.bnamericas.com/news/oilandgas/Petrobras_reaffirms_investments_despite_contract_changes; AP Worldstream, "Three more oil companies in Venezuela agree to state-controlled joint ventures," 6 October 2005; BBC News, "Venezuela gives Exxon Ultimatum," 6 October 2005, http://news.bbc.co.uk/2/hi/americas/4544390.stm.

148. USA Today, http://usatoday30.usatoday.com/money/economy/2007-06-26-2141258034_x.htm, June 26, 2007.

149. Ibid.

150. Herbert Smith Freehills, LLP, "ExxonMobil files ICSID Arbitration Proceedings Against Venezuela; Energy Investors in South America on Alert as more World Bank Departures Threatened," October 31, 2007, http://www.lexology.com/library/detail.aspx?g=af22d8b6-00d4-4509-aa68-579abb51c7fc.

151. International Centre for the Settlement of Investment Disputes, Case Details, ConocoPhillips Petrozuata B.V., ConocoPhillips Hamaca B.V. and ConocoPhillips Gulf of Paria B.V. v. Bolivarian Republic of Venezuela, (ICSID Case No. ARB/07/30) https://icsid.worldbank.org/ICSID/FrontServlet.

152. Leopoldo Olavarría, Natalija Vojvodic and Sergio Casinelli, "The Venezuelan State Reserves Assets and Services related to Hydrocarbon Primary Activities," May 2011, http://www.nortonrosefulbright.com/knowledge/publications/58825/the-venezuelan-state-reserves-assets-and-services-related-to-hydrocarbon-primary-activities.

153. Tamsin Carlisle, "Venezuela Seizes 60 firms," May 9, 2009, http://www.thenational.ae/business/energy/venezuela-seizes-60-firms.

154. Associated Press, "Venezuela Seizes Oil Rigs Owned by US Company," June 24, 2010, http://www.cnbc.com/id/.

155. Bloomberg, "Williams Files Arbitration Claim After Venezuelan Seizures," April 1, 2011, https://www.whatsnextvenezuela.com/tag/united-states/page/20/.

156. Timeline of Expropriations, "What's Next Venezuela," https://www.whatsnextvenezuela.com/media-kit/timeline-of-expropriations/.

157. Latin American Herald Tribune, "U.S. Driller Helmerich & Payne Sues Venezuela over Expropriated Rigs," June 25, 2014, http://laht.com/article.asp?CategoryId=10717&ArticleId=428308.

158. The Devil's Excrement, "ExxonMobil Versus PDVSA: Arbitration and Numbers," January 1, 2012, http://devilsexcrement.com/2012/01/02/exxonmobil-versus-pdvsa-arbitration-and-numbers/.

159. Ibid.

160. Ibid.

161. Ibid.

162. What's Next Venezuela, "Timeline of Expropriations," https://www.whatsnextvenezuela.com/media-kit/timeline-of-expropriations/.

163. Milbank, Tweed, Hadley and McCloy, LLP, "Venezuela Withdraws from the World Bank's International Centre for Investment Disputes, Litigation," January 30, 2012, http://www.milbank.com/images/content/7/2/7277/Venezuela-Withdraws-From-ICSID-1-30-2012.pdf.

164. What's Next Venezuela, "Timeline of Expropriations," https://www.whatsnextvenezuela.com/media-kit/timeline-of-expropriations/.

165. Weil, "2012–2013 Litigation Wins Report," http://www.weil.com/files/Publication/de949492-1eb8-476b-86f5-295b2dd17d6a/Presentation/PublicationAttachment/e02fcfe9-476f-4cb9-969c-3bb13d2379ae/2012_13_Litigation_Wins_Report.pdf.

166. Charles Wiley Armbrust, "Political Trends in Latin American Oil: Mexico and Venezuela," *Akin Gump, Strauss, Hauer and Feld LLP,* December 2, 2013, http://www.akingump.com/en/experience/industries/energy/speaking-energy/political-trends-in-latin-american-oil-mexico-and-venezuela-1.html.

167. Courthouse News Service, "Oil Firms May Have Case Against Venezuela," September 30, 2013, http://www.courthousenews.com/2013/09/30/61616.htm.

168. DW, "Venezuela Agrees to Pay ExxonMobil for Oil Nationalization," October 10, 2014, http://www.dw.de/venezuela-agrees-to-pay-exxonmobil-for-oil-nationalization/a-17985903.

169. BBC News, "Analysis: Latin America's New Left Axis," 18 April 2006, http://news.bbc.co.uk/2/hi/americas/4916270.stm.

170. BBC News, "Argentina to Expropriate Repsol Oil Subsidiary YPF," April 16, 2012, http://www.bbc.com/news/business-17732910.

171. ICSID, https://icsid.worldbank.org/apps/icsidweb/cases/Pages/casedetail.aspx?CaseNo=ARB/12/38.

172. Almudena Calatrava, "Problems with New Argentine Debt Payment Law," *The Associated Press*, September 13, 2014, http://www.usatoday.com/story/money/markets/2014/09/13/problems-seen-with-new-argentine-debt-payment-law/15580155/.

173. Daniel Bases, Richard Lough and Sarah Marsh, "Argentina, Lead Creditors Settle 14-year Debt Battle for $4.65 Billion," *Reuters*, http://www.reuters.com/article/us-argentina-debt-idUSKCN0W2249, March 1, 2016.

174. Hugh Bronstein and Sarah Marsh, "Argentina Returns to Global Debt Markets after 15-years," March 19, 2016, http://www.reuters.com/article/us-argentina-bonds-bids-idUSKCN0XG2W0.

175. Trading Economics, "Argentina's Government Budget Deficit," http://www.tradingeconomics.com/argentina/government-budget.

176. Trading Economics, "Venezuela Budget Deficit," http://www.tradingeconomics.com/venezuela/government-budget.

177. The Economist, "Venezuela: A Nation in a State," February 18, 2016, http://www.economist.com/blogs/graphicdetail/2016/02/graphics-political-and-economic-guide-venezuela.

178. Benjamin C. Esty, "Petrozuata: A Case Study of the Effective Use of Project Finance," *Bank of America Journal of Applied Corporate Finance*, Vol. 12, No.3, Fall 1999, https://faculty.fuqua.duke.edu/~charvey/Teaching/BA456_2002/Esty_JACF.pdf.

179. Alexander Osipovich, "TNK-BP Saga Raises Questions about BP's Handling of Political Risk," *Energy Risk,* March 19, 2013, http://www.risk.net/energy-risk/feature/2253578/tnkbp-saga-raises-questions-about-bps-handling-of-political-risk.

180. Erin E. Arvedlund, "A New Twist in Russia's Yukos Oil Affair," *The New York Times,* April 16, 2004, http://www.nytimes.com/2004/04/16/business/a-new-twist-in-russia-s-yukos-oil-affair.html?n=Top percent2fReference percent2fTimes percent-20Topics percent2fPeople percent2fL percent2fLebedev percent2c percent20Platon. David M. Woodruff. "Khodorkovsky's gamble: from business to politics in the YUKOS conflict," *LSE Research Online*. London School of Economics, 2003, http://eprints.lse.ac.uk/3689/1/Khodorkovsky percentE2 percent80 percent99s_gamble-from_business_to_politics_ percent28LSERO percent29.pdf.

181. Courtney Weaver, "Time Line: The Rise and Fall of Yukos," *Financial Times*, July 28, 2914, http://www.ft.com/intl/cms/s/0/f371c836-1645-11e4-93ec-00144feabdc0.html#axzz3O9da8H3P.

182. Alexander Osipovich, "TNK-BP Saga Raises Questions about BP's Handling of Political Risk," *Energy Risk,* March 19, 2013, http://www.risk.net/energy-risk/feature/2253578/tnkbp-saga-raises-questions-about-bps-handling-of-political-risk.

183. Miriam Elder, "Russia look to Control World's Gas Prices," *Telegraph,* December 27, 2008, http://www.telegraph.co.uk/earth/energy/gas/3982543/Russia-look-to-control-worlds-gas-prices.html; Tom Parfitt, "Kremlin Attack Dog Vows to Take on Shell in the Battle of Sakhalin," *The Guardian,* October 4, 2006, http://www.theguardian.com/business/2006/oct/04/russia.oilandpetrol; Melissa Kite and Nicholas Holdsworth, "Russian Bullying Over Oil is 'a wake-up call," *Telegraph*, December 24, 2006, http://www.telegraph.co.uk/news/uknews/1537759/Russian-bullying-over-oil-is-a-wake-up-call.html.

184. The Telegraph, "TNK-BP: A Troubled History," July 30, 2014, http://www.telegraph.co.uk/finance/newsbysector/energy/oilandgas/9305157/TNK-BP-a-troubled-history.html.

185. Terry Macalister, "Oligarchs to sue TNK-BP after Failing to Agree Control of Company," *The Guardian,* June 11, 2008, http://www.theguardian.com/business/2008/jun/12/bp.oil1?gusrc=rss&feed=business.

186. Alexander Osipovich, "TNK-BP Saga Raises Questions about BP's Handling of Political Risk," *Energy Risk,* March 19, 2013, http://www.risk.net/energy-risk/feature/2253578/tnkbp-saga-raises-questions-about-bps-handling-of-political-risk.

187. The Telegraph, "TNK-BP: A Troubled History," July 30, 2014, http://www.telegraph.co.uk/finance/newsbysector/energy/oilandgas/9305157/TNK-BP-a-troubled-history.html.

188. Alexander Osipovich, "TNK-BP Saga Raises Questions about BP's Handling of Political Risk," *Energy Risk,* March 19, 2013, http://www.risk.net/energy-risk/feature/2253578/tnkbp-saga-raises-questions-about-bps-handling-of-political-risk.

189. Jennifer Wishon, "BP Replaces CEO Hayward, reports $17 billion loss," *CBN News,* July 27, 2010, http://www.cbn.com/cbnnews/us/2010/July/BP-Replaces-Hayward-Reports-17-Billion-Loss-/.

190. Shamil Yenikeyeff, "BP, Russian Billionaires, and the Kremlin: A Power Triangle that Never Was," Oxford Energy Comment, November 2011, http://www.

oxfordenergy.org/wpcms/wp-content/uploads/2011/11/BP-Russian-billionaires-and-the-Kremlin.pdf.

191. Alexander Osipovich, "TNK-BP Saga Raises Questions about BP's Handling of Political Risk," *Energy Risk*, March 19, 2013, http://www.risk.net/energy-risk/feature/2253578/tnkbp-saga-raises-questions-about-bps-handling-of-political-risk.

192. Vladimir Soldatkin and Andrew Callus, "Rosneft Pays Out in Historic TNK-BP Deal Completion," *Reuters*, March 22, 2013, http://uk.reuters.com/article/2013/03/22/uk-rosneft-tnkbp-deal-idUKBRE92K0IX20130322.

193. Alexander Osipovich, "TNK-BP Saga Raises Questions about BP's Handling of Political Risk," *Energy Risk*, March 19, 2013, http://www.risk.net/energy-risk/feature/2253578/tnkbp-saga-raises-questions-about-bps-handling-of-political-risk.

194. Ibid.

195. Ibid.

Summary and Conclusion

The introduction explained why evaluating and predicting political risk is difficult. First, changes in government policy and political regimes occur with sufficient frequency that it is inevitable that some oil and gas investments will be made in developing countries at a time when national policy is favorable to private investment, but later subject to policies and laws that are unfavorable. Second, there is often a shift in bargaining power over the life of a project between the nation supplying the natural resource and the multinational companies providing the financial capital and technical expertise, even in the absence of a change in policy or political regime. Third, the large amount of the financial and physical capital required in oil and gas exploration and production; and the immobility of these assets make them particularly vulnerable to the obsolescing bargain.

The effectiveness with which international oil companies (IOCs) manage political and investment risk is important to the firm and to the global economy for several reasons. First, oil and gas are essential for sustaining current levels of economic activity and for making economic growth possible. Second, the balance between supply and demand determines the price of oil and natural gas; and the price of these commodities affects the economies of all nations. However, when the risk of contract renegotiation or expropriation appears to be increasing, investments in oil and natural gas that would otherwise be made will not be made, thereby reducing the supply and increasing the price of oil and natural gas. Third, the adequacy of oil and gas investment is a matter of national security, which affects the political, military and economic relations between countries.

Chapter 1 began by defining political risk "as the impact of politics on markets. Political risk is influenced by the passage of laws, the foibles of leaders, and the rise of popular movements, in short, all the factors that might politically stabilize or destabilize a country."[1] This book has focused on above-ground risk, which includes adverse changes in economic conditions, either intentional or unintentional; the inability to enforce the terms of a contract; changed fiscal terms imposed by the host government; resource nationalism; discriminatory regulatory behavior; and the placement of geopolitical considerations (national security, energy security and political alliances) above legal and financial commitments.[2]

Next, chapter 1 examined the difference between international portfolio investments (debt) and foreign direct investment (equity). National governments acting through the executive branch or the central bank are more likely to intervene in matters related to international debt markets than they are in matters related to foreign direct investment for two reasons.[3] Multinational corporations have mostly independent interests in investment disputes, making the disputes between investors and host governments separable issues, but financial organizations have joint interests because of the network of interbank liabilities between them.

Next, chapter 1 examined some of the theoretical frameworks that have been developed to explain the operation of markets and the role that institutions play in making those markets more efficient. In its classical economic formulation, markets are assumed to be efficient even in the absence of regulation. In its neoclassical formulation, markets are thought to work efficiently most of the time, but may require minimal regulation to maintain an orderly market. A third approach to economics in general and political economy specifically is the "new institutional economics" (NIE). The approach differs from neoclassical economics in several important ways. First, it rejects the assumption that people or companies act independently on full and relevant information. Second, NIE recognizes that engaging in transactions involves the incurrence of transaction costs and transaction risk, which influence the decision-making process and therefore the decision. Third, the two preceding points make a formal system of rules desirable because formal rules lower information and transaction costs and reduce the risk associated with a transaction. In this approach, the two principal classes of participants are institutions and organizations. *Institutions* are the "rules of the game", consisting of both the formal legal rules and the informal social norms that govern individual behavior and structure social interactions. *Organizations* are those groups of people and the governance arrangements they create to coordinate their group action against other groups also performing as organizations, for example, business firms, universities, clubs, medical associations and unions.[4]

Institutions (contracts, domestic laws, treaties, international law, formal rules and informal customs) and the entities that implement them (regulatory agencies, financial exchanges, nongovernmental agencies and intergovernmental agencies) are involved in virtually every stage of the oil and gas exploration, development and production process. This involvement extends beyond the neoclassical assumptions regarding rules and regulations needed to maintain an orderly market and includes laws regarding capital formation, ownership structure, foreign direct investment, capital allocation, domestic and international product pricing, taxation, capital repatriation and social service expenditures.

Chapter 2 began by examining the politics, policies and objectives of some of the larger oil-exporting countries and their governments. National governments are the representatives of the people and therefore, at least in theory, act to enhance the welfare of the people in that country. In countries exporting oil and natural gas, this principally consists of the development of energy policies related to the discovery, production and sale of oil and gas. Energy policy includes approving the organizations that will be granted drilling rights in the country, the rate of production allowed, and the division of revenue between the national government and the IOCs or foreign national oil companies (NOCs) operating in the country. The energy policy of national governments can change over time and this represents in the broadest sense, the risk to foreign direct investment. The national governments examined in this section included Saudi Arabia, Russia, Venezuela, Mexico, China, Nigeria and Iraq.

Next, chapter 2 reviewed the economic forces leading to the development of NOC, the formation of their policies and their behavior in general. Victor identified four reasons why governments have created and maintained NOCs rather than authorize the creation of private sector companies and simply impose taxes on them.[5] The first is related to the *weakness of public institutions* in many oil-producing countries. Governments must perform many functions, including the regulation of a complex oil and gas industry. The institutions needed to regulate the industry are often weak or nonexistent in developing countries. Governments may create a NOC to improve the collection of information and the regulation of the oil and gas industry by giving the government direct access to the country's oil and gas operations.[6, 7]

The second derives from the *inadequacy of the domestic private sector.* The suppliers and stages of production needed to explore for and produce oil and gas do not exist in many developing countries. Discovering previously unknown hydrocarbon resources and maximizing the value of the country's oil and gas resources requires coordination among many suppliers and stages of production. Governments often create a national oil company to help encourage the orderly development of these suppliers and stages of production in the country's oil and gas industry.[8]

A third reason is the *desire for internal control over the revenue created by oil and gas production.* Control over the cash inflow from the production and sale of oil and gas is important to a host government. In countries, whose governments primarily depend on the revenue derived from the export of oil and gas to finance their national budgets, one way to control the revenue is to control the NOC.[9]

The fourth is the *desire to project economic and political power abroad.* A national oil company can be used to exert political influence abroad as well as at home by threatening to cut off the supply of oil or gas to a country or countries; by reducing production of oil and gas to drive the price higher; or by increasing production to lower prices, putting economic pressure on other oil countries.

Studies of the political behavior of NOCs can be divided into two broad categories. One category emphasizes the principal-agent relationship between the host government and the NOC, in which the principal (the state and its people) try to encourage an agent (the NOC) to align its behavior with the principal's interests.[10] Another group of studies demonstrate how NOCs become political actors to advance their own agenda and increase their power. [11]

Chapter 2 also examined the changing role of IOCs. Between 1995 and 2005, the role of the IOCs evolved from technically integrated and self-sufficient exploration and development companies to general contractors that coordinate the operations of many suppliers and subcontractors who perform specific tasks. The IOCs also provide the project management and organizational skills these large projects require. The largest IOCs also function as bankers by providing or arranging the financing needed to explore and develop large fields in difficult environments.[12]

IOCs are not primarily political actors, but rather economic actors. They tend to be involved in political issues only to the extent that the actions of governments affect the economics of their operations and their future profitability. However, the price and the availability of oil and natural gas are of intense interest to governments and their citizens in developed and developing countries. Consequently, oil and natural gas are the object of political discourse and policy making at the state and federal level.

Chapter 3 began by summarizing the provisions most often included in oil or gas contracts to minimize disputes and resolve those that occur. These include a general clause that requires both parties to act "in good faith", "use reasonable efforts" and fulfill their obligations "in a timely manner". Other provisions include an equilibrium clause intended to compensate for changes in the legal or the investment environment in which a contract was originally signed; and clauses specifying the choice of law and the court or tribunal to be used in the resolution of disputes.

Next, chapter 3 examined the relationship between domestic courts, foreign courts and arbitration tribunals. Most foreign investors insist on a clause specifying that disputes are to be resolved through an arbitration tribunal rather than a domestic court in the host country. Third, chapter 3 examined the most important clauses in bilateral investment treaties and the problems that arise in their interpretation. The most common provisions include a statement of purpose, choice of law, standards of treatment and the legal instruments to be applied in a dispute. The most contentious issues in cases brought before an international court or tribunal are the interpretation of the terms "fair and equitable treatment," "fair and equitable compensation," "minimum standard" and the rules governing expropriation of assets.

Fourth, most disputes are the result of a disagreement over the interpretation of specific terms and conditions in a contract; developments that were not foreseen when the contract was prepared; or changes in the host government's policies. Possible reasons for bringing a dispute before an international court or tribunal include the failure of the contract to provide for the specific contingency that has arisen; repudiation of a contract by the host state; a lack of confidence in the fairness of the domestic courts in the host country; and the inclusion of a provision in the contract that specifies that an international court or tribunal is to be used.

Fifth, the number of bilateral treaties a country has signed and the quality of governance in a country are not reliable predictors of the number of cases that will be brought against the country in an international court or tribunal. Sixth, chapter 3 evaluated the fairness of the decisions reached by international tribunals. International courts or tribunals decided in favor of the investor in 60 percent of cases and in favor of the state in 40 percent of cases. In twenty-five of twenty-eight cases, the investors were awarded less than the amount they requested. This is in part, because claimants usually seek compensation equal to the fair market value of the assets expropriated or the market value of the breach of contract, but courts are more likely to award an amount equal to the book value, even when they decide in favor of the claimant.

Chapter 3 also examined the efficiency of international courts and tribunals. One measure of the efficiency is the time from initiation of the case to the date the court or tribunal rendered its decision. The average time for a decision from an international court or tribunal was 4.6 years. The average time to settlement outside a court or tribunal was 2.6 years. The average time in adjudication for those cases still pending is 3.6 years.

Seventh, Table 3.2 demonstrated that the expropriation of oil and gas assets had already begun in the 1960s (twenty), but increased substantially between 1971 and 1979 (sixty-nine). Expropriations nearly ceased between 1980 and 2003 and increased again between 2004 and 2011. Table 3.2 also supports the conclusion that the number of expropriations increases when the rate of change in the price of oil increases and the number of expropriations decreases when the rate of change in the price of oil decreases.[13]

Chapter 4 examined the principal sources of financing in the oil and gas industry, financing structures, fiscal regimes, alliances, management of commercial and noncommercial risk, financial reporting, operational transparency, financial reporting quality and the estimation and valuation of oil and gas reserves. First, commercial bank debt is the principal source of financing in the oil and gas industry, followed by bonds, project financing and equity. Master limited partnerships, venture capital and multilateral financing play a smaller role. Second, the centralized finance system is still the most common financing structure in the oil and gas industry. Other financing structures include joint ventures, project financing, structured project financing and special purpose entities.

Third, there are three fiscal regimes for controlling assets and the distribution of revenue in the oil and gas industry: (1) a concession or lease agreement (sometimes referred to as a royalty/tax agreement); (2) a production-sharing agreement or production-sharing contract; and (3) a service contract. A host country must balance two competing objectives. The first is to encourage investment by international oil and gas exploration companies by providing sufficient incentive for oil and gas companies to invest in the country. The second is to balance the interests of the IOCs with the interests of its citizens and export customers.[14]

Fourth, alliances between NOCs and IOCs can increase the efficiency of the international oil and gas industry because each has unique and complementary capabilities. NOCs can manage resource and production controls in a manner that the host government perceives to be in the national interest. IOCs possess substantial technical expertise, project management skills and the ability to define and develop new products and markets.[15]

Fifth, after a project is in commercial operation, it still faces several commercial and noncommercial risks. The *commercial risks* include: (1) a decrease in revenue as a result of an unexpected decline in product demand or an increase in aggregate product supply that reduces the price of oil and natural gas; (2) an unexpected increase in operating cost; (3) a loss of revenue from business interruption or lack of business continuity; (4) property damage; (5) labor disputes and local labor management problems; and (6) liability (workers' compensation, product liability, environmental liability and general liability). In addition, oil and gas companies operating onshore or offshore the United States are required to demonstrate that they have sufficient financial resources to pay for damages and the cost of cleanup created by an oil spill.

Noncommercial risks can be divided into two broad categories (financial risk and political risk). The financial risks include changes in interest rates, exchange rates, commodity prices and inflation. Political risk includes: (1) asset impairment (degradation, destruction and expropriation); (2) political violence and war; (3) insecure property rights (contract frustration or abrogation, patent violations, wrongful calling of guarantees, host country failure to honor guarantees and changes in host country laws); and (4) currency inconvertibility and capital flow restrictions.

There is a wide variety of legal instruments available for allocating and managing commercial risk. These instruments include property and casualty insurance, reinsurance, environmental liability insurance, the U.S. Oil Spill Liability Trust Fund and self-insurance. The instruments for managing noncommercial risk include commodity futures, forward contracts, options, swaps, long-term purchase agreements and political risk insurance.

Several empirical studies have concluded that higher-quality financial reporting increases the efficiency of capital markets.[16] These studies evaluated the quality of financial reporting and market efficiency in buying and selling financial securities on an organized exchange. They did not relate the quality of financial reporting and market efficiency to foreign direct investment. It cannot be assumed that what is true in organized financial markets and exchanges is true of foreign direct investment.

Estimating and valuing reserves is the most difficult reporting challenge because future production can only be approximated given the technical, geologic and pricing uncertainty surrounding oil and natural gas, but the estimated level of oil and gas reserves effects the price and supply of oil and gas. In the short term, current oil inventories, that is, oil above ground, is the most important determinant of current oil prices and the current level of capital expenditures made in the search for new oil and gas reserves. However, in the long term, estimates of proved and probable reserves below ground play the most important role in determining the price of oil and the level of capital expenditures.[17]

Chapter 5 noted that there are no internationally agreed-upon set of rules for trading energy resources and energy investments, but the absence of a global energy investment agreement or even a substantial number of multilateral agreements may not be important, because a relatively small number of disputes are adjudicated by reference to any investment treaty. Second, the outcome of an international court proceeding or tribunal is dependent on the standard of treatment applied and the definition of fair and equitable compensation by the court or tribunal. Therefore, even if a global investment treaty existed, the problems of interpretation would still exist.

The oil and gas community comprised of national governments, national oil companies and IOCs have created some institutions for exchanging information. The formation of the International Energy Forum (IEF) has led to more cooperation on energy issues among nations, for example the Joint Oil Data Initiative and Joint Gas Data Initiative, but these organizations have been successful primarily in compiling historical information. Obtaining data regarding the location and level of future capital expenditures has not been possible so far.

Since 1990, the International Financial Reporting Standards (IFRS), U.S. Generally Accepted Accounting Principles (GAAP) and Japanese GAAP have continued to converge and is an example of organizations (investors, banks and companies) reshaping the institutional environment to increase the efficiency of business transactions and capital allocation. The widespread adoption of IFRS since the 1990s has had a significant impact on financial reporting, capital markets, companies listed on public exchanges, and investors. Companies listed on public exchanges can raise financial capital in other countries without having to first restate their financial results using that country's accounting standards.

As the number of conventional drilling prospects decreased, the value of signature bonuses has increased. In the beginning, signature bonuses were between $1 million and $10 million per bid, however, recent exploration and development agreements have included signature bonuses exceeding $100 million.[18] The size of these signature bonuses and the occasional diversion of these payments by government officials to personal bank accounts have produced an ongoing controversy about their impact on host country governance and corruption. On June 27, 2016, the Securities and Exchange Commission (SEC) issued a rule requiring oil, gas and mining companies listed on U.S. stock exchanges to disclose the payments they make to the U.S. and foreign governments.[19]

Energy diplomacy has not lived up to its promise, in part because oil inventories and production are at high levels, therefore diminishing the value of long-term sources of supply. However, when inventory levels return to lower levels and demand exceeds supply the economic benefits of energy diplomacy may be more evident.

Every country has rules regarding foreign direct investment. These rules serve a variety of purposes, including economic stabilization, economic development, technology transfer and national security. The small number of instances in which the U.S. government has intervened in a proposed foreign direct investment in the oil and gas sector or any other sector of the U.S. economy indicates that in the United States foreign direct investment is not significantly impeded by government restrictions.

Several conclusions follow implicitly or explicitly from the previous chapters. First, investment disputes in the international oil and gas industry are governed by treaties, international customs, general principles of law as recognized by civilized nations, the decisions of national and lower courts, scholarly writings and to a lesser extent by domestic legislation. Second, the United States is a party to forty-seven bilateral investment treaties. Forty of these bilateral treaties are currently in force, six are signed, but not in force, and one is terminated. However, the United States is a party to another sixty-seven treaties that contain investment provisions.[20] There were 196 countries in the world at the time of writing, so the United States has either a bilateral treaty or a treaty with investment provisions with approximately 58 percent of the countries of the world.

Third, in twenty-six of the sixty-three cases examined in this book, U.S. oil and gas companies operating in other countries were the claimant; and in none of the sixty-three cases was the United States named as the respondent (defendant). U.S. oil and gas companies therefore accounted for 41 percent of the investment disputes related to oil and gas exploration, production and transportation brought before an international court or tribunal. Although this is a large portion of the oil and gas investment disputes, it still represents a small percentage of the oil and gas investments made between 1996 and 2015.

The institutions available for the resolution of investment disputes can be arranged in a hierarchy based on their relative importance. The most important institutional mechanisms used by international oil and natural gas companies are contracts, informal and interpersonal relationships and transaction transparency between the IOC and the host government. The second most important institutions include investment treaties, customary international law and international courts and tribunals. A third line of defense includes multilateral treaties and international forums. However, only a small number of these institutions are the direct result of developments in the oil and gas industry. Most of these institutions have developed in response to the needs of foreign direct investors in general and have then been adapted for use in the oil and gas industry.

For the company that finds itself in a dispute with a foreign government over contract rights or ownership, the experience is frustrating and unpleasant. However, the evidence suggests that the international legal system available for the resolution of disputes is for the most part reasonably fair and effective. This conclusion obviates the need for significant or radical changes in the international legal system, except for more effective enforcement mechanisms for international court and arbitral decisions. However, the sovereignty of individual nations will always impose a constraint on the expansion and strength of international enforcement mechanisms.

The evidence in this book is consistent with North's observation that "incremental change [in institutions] comes from the perceptions of the entrepreneurs in political and economic organizations that they [the entrepreneurs] could do better by altering the existing institutional framework at some margin" [21] and that "institutions do not emerge spontaneously to create and nurture the market, but rather reflect the interests of those players, in a position, to put them in place." [22] In other words, oil and gas companies will act through industry associations, paid policy advocates and their own senior executives to influence government policy and institutions at the state, national and international level when the need arises.

Finally, Columbia University law professor Louis Henkin has observed "It is probably the case that almost all nations observe almost all principles of international law and almost all of their obligations almost all of the time."[23] Henkin was referring primarily to the relations between nation states, that is, public law, but the evidence in this book suggests that this observation is also true of international commercial law, that is, private law. This is the case despite the fact, that there is no supranational organization capable of enforcing international contracts and bilateral treaties. This is in part because parties to an agreement want the other side to comply with its obligations and one way to encourage this compliance is to meet one's own obligations; and second, parties honor their commitments because if they do not, other parties will be reluctant to contract with them again.

NOTES

1. Ian Bremmer, "Managing Risk in an Unstable World," *Harvard Business Review*, June, 2005.

2. Ibid., p. 239–240.

3. Charles Lipson, *Standing Guard – Protecting Foreign Capital in the Nineteenth and Twentieth Centuries.* (Los Angeles, CA: University of California Press, 1985), p. 170.

4. Douglass C. North, "Institutions," *Journal of Economic Perspectives*, Vol. 5, No. 1, Winter, 1991, pp. 97–112.

5. David G. Victor, "National Oil Companies and the Future of the Oil Industry," *The Annual Review of Resource Economics*, Vol. 5, June, 2013, p. 445–462.

6. L. E. Grayson, *National Oil Companies.* (Chichester, UK: Wiley, 1981).

7. C. McPherson, *National Oil Companies: Ensuring Benefits and Dealing with Systemic Risks.* In *The Handbook of Global Energy Policy*, ed. A. Goldthau, (Hoboken, NJ: Wiley-Blackwell, 2013), pp. 146–156.

8. C. van der Linde, *The State and the International Oil Market: Competition and the Changing Ownership of Crude Oil Assets.* (Boston: Kluwer Academic Publishers, 2000).

9. David G. Victor, "National Oil Companies and the Future of the Oil Industry," *The Annual Review of Resource Economics*, Vol. 5, June, 2013, pp. 445–462.

10. G. J. Miller, "The Political Evolution of Principal-Agent Models," *Annual Review of Political Science,* Vol. 8, No. 1, 2005, pp. 203–225.

11. B. Mommer, *Global Oil and the Nation State.* (UK/New York: Oxford Univ. Press, 2002.

12. Amy Myers-Jaffe and Ronald Soligo, "The International Oil Companies," The James A. Baker III Institute for Public Policy, November 2007, pp. 36–37, http://bakerinstitute.org/files/707/.

13. Sergei Guriev, Anton Kolotilin and Konstantin Sonin, "Determinants of Nationalization in the Oil Sector: A Theory and Evidence from Panel Data, *Journal of Law, Economics & Organization*, Vol. 27, No. 2, 2011, p. 301.

14. Andrew Inkpen and Michael H. Moffett, *The Global Oil and Gas Industry—Management, Strategy and Finance.* Chapter 6 (Tulsa, OK: PennWell Corporation, 2011), p. 217.

15. Robert A. James, *Strategic Alliances between National and International Oil Companies.* (Working paper #104, Freeman Spogli Institute for International Studies, October, 2011), p. 12.

16. Gary C. Biddle, Gilles Hilary, and Rodrigo S. Verdi, "How Does Financial Reporting Quality Relate to Investment Efficiency?" *Journal of Accounting and Economics,* Vol. 48, No. 2–3, 2009, p. 112–131; R. Bushman and A. Smith, "Financial Accounting Information and Corporate Governance," *Journal of Accounting and Economics,* Vol. 31, 2001, pp. 237–333; P. Healy and K. Palepu, "Information Asymmetry, Corporate Disclosure, and the Capital Markets: A Review of the Empirical Disclosure Literature," *Journal of Accounting and Economics,* Vol. 31, pp. 405–440; R. Lambert, C. Leuz, and R. Verrecchia, "Accounting Information, Disclosure, and the cost of Capital," *Journal of Accounting Research*, Vo. 45, 2007, pp. 385–420; G. Biddle and G. Hilary. "Accounting Quality and Firm-level Capital Investment," *The Accounting Review*, Vol. 81, 2006, pp. 963–982.

17. Klaus Mohn and Petter Osmundsen. "Asymmetry and Uncertainty in Capital Formation: An Application to Oil Investment," *Applied Economics*, June 2011, pp. 4387–4401.

18. Andrew Inkpen and Michael H. Moffett, *The Global Oil and Gas Industry—Management, Strategy and Finance.* (Tulsa, OK: PennWell Corporation, 2011), p. 228.

19. Marco Simons, "SEC Issues Long-Awaited Transparency Rule for Oil, Gas and Mining," EarthRights International, https://www.earthrights.org/media/sec-issues-long-awaited-transparency-rule-oil-gas-and-mining.

20. UNCTAD, International Investment Agreements, http://investmentpolicyhub.unctad.org/IIA/AdvancedSearchBITResults.

21. Douglass C. North, *Institutions, Institutional Change and Economic Performance.* (Cambridge, UK: Cambridge University Press, 1990), p. 8.

22. Ibid., pp. 89–90.

23. Louis Henkin, *How Nations Behave.* (New York: Columbia University Press, 1979) p. 47.

Appendix I

CRUDE OIL PRICES 1861–2015

Figure I.1 presents crude oil prices in the money of the day and in 2015 dollars.

For a compilation of Petroleum Liquids Consumption, Production and Reserves by Country see https://www.bp.com/content/dam/bp/pdf/energy-economics/statistical-review-2016/bp-statistical-review-of-world-energy-2016-full-report.pdf.

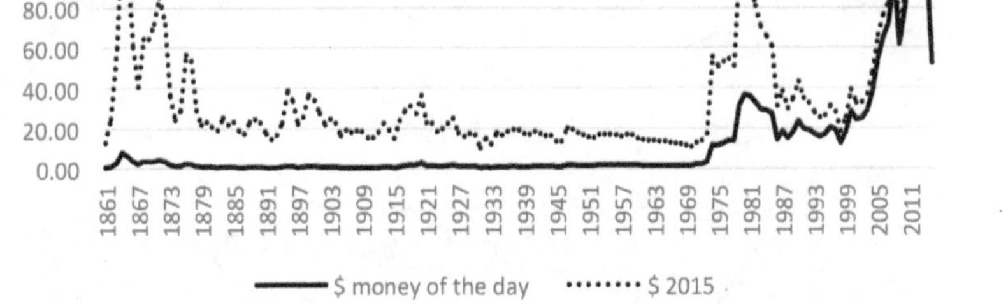

Figure I.1 Crude Oil Prices 1861–2015. *Source*: Adapted from BP, Statistical Review of World Energy – Data Workbook. June 2016. http://www.bp.com/en/global/corporate/energy-economics/statistical-review-of-world-energy/downloads.html.

Appendix II

EXPROPRIATIONS AND CHANGES IN THE TREND OF OIL PRICES

Guriev, Kolotilin and Sonin analyzed the determinants of oil and gas asset nationalizations (expropriations) from 1960 to 2006. They found that the largest number of nationalizations took place when oil prices were rising rapidly relative to the long-term price trend; and were more likely to occur in countries in which political institutions were weak.[1] For a more complete presentation of their results and conclusion see:

Guriev, Sergei, Anton Kolotilin and Konstantin Sonin, "Determinants of Nationalization in the Oil Sector: A Theory and Evidence from Panel Data," *Journal of Law, Economics & Organization*, 27 (2) (2011) 301

A DETAILED CHRONOLOGY OF PRIVATIZATIONS AND EXPROPRIATIONS (1989 TO 2014)

1989—Repsol (a Spanish oil company) was privatized by selling 100 percent of the company to the public between 1989 and 1997.[2]

1990–1991—No expropriations were reported.

1992—Total (a French oil company) began the process of privatizing the company. The company sold a 30 percent interest to the public in 1992 and the remaining 70 percent to the public in 1998.[3]

08/24/1992—ARCO and Sun Co. Inc. agreed to separate settlements totaling almost $261 million that resolved their claims over oil field assets expropriated by Iran in 1978–1980. The agreements were subject to approval by the Iran-U.S. claims tribunal at The Hague. The tribunal was set up in 1981 to resolve foreign claims to assets nationalized by the government of Ayatollah Khomeini after the fall of the Shah of Iran during the 1978–79 Iranian revolution.[4]

12/13/1993—(1) Effective November 18, 1993, Peru established a new state petroleum company, Perupetro SA, to oversee new and existing contract responsibilities of Petroleos del Peru SA (Petroperu), which was to be privatized in 1994. Petroperu would continue to operate its oil fields, refining, transportation and marketing assets. (This privatization had not occurred as of August 16, 2016.) (2) In addition, the government of Peru made a $30 million payment to the American International Group (AIG) in late September 1993. AIG was owed $184.8 million, because they were Belco's insurer and Belco's assets had been expropriated by the Peruvian government in December 1985. [5, 6]

1993—Oil and Natural Gas Corporation Limited (ONGC) sold a 16 percent interest to the public.[7]

1994—OAO Gazprom (a Russian natural gas company) sold a 62 percent interest to the public; and Lukoil (a Russian oil company) became a private sector company (100 percent).[8]

09/05/1994—Peru agreed to pay $55 million in compensation to a German group for the military government's 1974 expropriation of the undeveloped *Aguaytia* gas field and the producing *Maquia* oil field in Peru's central jungle.[9]

1995—(1) Three privatizations occurred. IndianOil Corporation Limited sold 11 percent of its shares to the public. As of 2014, the Indian government owned 79 percent and the public owned 21 percent. (2) ENI (an Italian company) sold a 15 percent interest in the company to the public. (3) Petróleo Brasileiro S.A. sold a 49 percent interest in the company to the public.[10]

1996–1999—No significant privatizations or expropriations occurred.

2000—Two significant partial privatizations occurred. China Petrochemical Corporation (SINOPEC) began the sale of a 24 percent interest in the company to the public; and PetroChina Company Limited (PetroChina) began the sale of a 14 percent interest in the company to the public.[11]

2001—Three significant partial privatizations occurred. (1) The Chinese National Offshore Oil Corporation sold a 29 percent share of its common stock to the public. (2) Statoil, a Norwegian national oil company, sold a 38 percent interest to the public. (3) Public ownership in ENI was increased to 70 percent.[12]

2002—No expropriations were reported.

2003—Following a tax reassessment, the Russian government presented OAO Yukos with a series of tax claims that totaled US $27 billion. Yukos' assets were frozen by the government at the same time and the company was therefore unable to pay these tax claims.[13] Most of Yukos' assets were subsequently sold at low prices to oil companies owned by the Russian government.[14] (See entry for 11/19/2004.) The Parliamentary Assembly of the Council of Europe condemned Russia's campaign against Yukos and its owners, as "manufactured" for political reasons and a violation of human rights.[15]

03/02/2004—Marathon Oil Corp. cancelled its proposed Liquefied Natural Gas (LNG) complex in Baja California, after the state government appropriated the land near Tijuana known as "Il Monumento," for public use, which included the land Marathon had selected for its project. Marathon had options to purchase the land, but did not own it when the property was taken.[16]

07/20/2004—Bolivian President, Carlos Mesa, declared victory regarding the referendum to increase state involvement in the country's gas reserves, after early returns showed that Bolivians approved all five measures on the ballot.[17]

08/02/2004—The Overseas Private Investment Corp denied charges made by Venezuela's state-owned oil company Petróleos de Venezuela S.A., that the U.S. agency's decision to pay $6 million on an insurance claim to Science Applications International Corp. , an information technology contractor, was politically motivated.[18]

11/19/2004—The Russian government announced bidding opened for an auction of a subsidiary of OAO Yukos to be held on December 19, 2004. The auction of 76.79 percent of *Yuganskneftegaz* was open to international companies. The Russian government required a $1.7 billion deposit to bid and set the minimum bid at $8.6 billion. Yukos claimed the subsidiary's value was $20 billion.[19]

2005—There were no expropriations or threats of expropriation.

05/08/2006—Bolivia demanded that Brazil's state-run Petroleo Brasileiro SA(Petrobras), Royal Dutch Shell PLC and other private firms turn over their Bolivian retail networks to state-owned Yacimientos Petroliferas Fiscales Bolivianos (YPFB) within 180 days.[20]

05/17/2006—Occidental Petroleum Corp. filed an arbitration claim against Ecuador on May 17, 2006, seeking reparation for losses following Ecuador's termination of Occidental's exploration and development contract and the immediate confiscation of the company's Amazon oil field operations in Block 15 and its Eden-Yuturi, Limonchcha, Indillana, Paca Norte, Paca Sur and Yanaquincha fields on May 15, 2006.

Occidental filed a claim with the International Centre for the Settlement of Investment Disputes, invoking the US-Ecuador Bilateral Investment Treaty to try to restore the company's rights in Ecuador and prevent the Ecuadorian government from turning operations over to a third party, until the claim was settled.[21]

06/05/2006—(1) Bolivia's President Evo Morales issued a decree reaffirming that Brazil's state-run Petrobras, Royal Dutch Shell PLC, and other private firms must turn over their Bolivian retail networks within a month to state-owned YPFB. He also demanded that Petrobras, which owned 25 percent of the country's retail outlets, give up its 27,250 barrel per day-refinery at Cochabamba and a 20,000 barrel per day-refinery at Santa Cruz de la Sierra.[22] (2)

Alaska's Governor Frank H. Murkowski released a revised natural gas pipeline agreement negotiated with the three North Slope oil producers, which he said the oil companies were ready to sign, if the state legislature approved it.[23]

08/01/2006—A Russian court declared OAO Yukos bankrupt.[24]

09/25/2006—(1) A Russian government injunction rescinded permits for the second phase of Sakhalin-2, alleging threats to marine life resulting from inadequate environmental safeguards. Shell and its partners Mitsubishi and Mitsui, disputed the claims, but revocation of the permits effectively suspended the project. In addition, Interfax reported that an official of the Natural Resources Ministry had implied that the other projects at Sakhalin Island were in violation of their licenses. At the same time, *Gazprom* was lobbying for a stake in the Sakhalin-2 venture. (2) The Financial Times reported threats by Russian prosecutors to suspend the license of TNK-BP to develop *Kovytka* gas field in eastern Siberia. In that case, as with Sakhalin-2, the allegations were related to environmental issues.[25]

01/15/2007—An agreement was signed between Royal Dutch Shell PLC and *OAO Gazprom*, the Russian state-owned natural gas company, under which *Gazprom* would become the majority shareholder in the Sakhalin-2 LNG Project. Under the terms of the agreement, *Gazprom* would purchase 50 percent plus one share of the project for $7.45 billion, thereby forcing Shell and its partners, Mitsui & Co. Ltd. and Mitsubishi Corp. to dilute their interest by 50 percent to accommodate their new partner, *Gazprom*. Shell was forced to give up majority control in one of its most valuable assets after having invested over $6 billion to develop the project.

Immediately following the agreement, Russian President Vladimir Putin held a press conference at which he announced that the environmental violations at Sakhalin-2 had been resolved to the satisfaction of the Russian regulatory authorities.[26]

06/27/2007—Petroleos de Venezuela S.A. (PDVSA) assumed ownership of ConocoPhillips's interests in the Petrozuata and Hamaca heavy-oil ventures in Venezuela and the offshore Corocoro development project, after PDVSA and ConocoPhillips were unable to reach an agreement under which ConocoPhillips would transition to an "Empresa Mixta" structure mandated by Venezuelan law. ConocoPhillips said it intended to record a complete impairment of its interest in its oil projects in Venezuela which it valued at $4.5 billion. Prior to the expropriation of its interests, ConocoPhillips held a 50.1 percent interest in Petrozuata, a 40 percent interest in Hamaca and a 32.5 percent interest in Corocoro.[27]

10/2007—Ecuador's President Rafael Correa imposed a windfall profits tax on oil operations whereby the government would receive 99 percent of oil profits, changing the prior law which required a 50–50 split of profits.[28]

2008—There were no reports of expropriation or privatization.

04/19/2010—Rafael Correa, president of Ecuador, the smallest oil producer within the Organization of Petroleum Exporting Countries, said a bill would be introduced in the legislature that would allow expropriation of those companies that refused to convert their production-sharing contracts into service contracts.[29]

07/02/2010—Venezuela nationalized eleven oil rigs owned by Helmerich & Payne (H&P).[30]

03/29/2011—Madagascar Oil of Houston, declared *force majeure* under the four production-sharing contracts for Blocks 3104, 3105, 3106 and 3107 that it operated in Madagascar to safeguard its rights under those agreements. The company said the declaration of *force majeure* was made in response to the threat of expropriation made by the minister of mines and hydrocarbons and the failure by the ministry to instruct the state regulatory authority to proceed with the approval of Madagascar Oil's 2011 work program.[31]

04/23/2012—Argentina took control of the energy company Yacimientos Petroliferos Fiscales, SA (YPF SA). The Spanish company Repsol, part owner of YPF SA issued a statement saying it would take all legal measures to preserve the value of its assets and the interests of its shareholders. Repsol valued YPF SA at $18 billion of which Repsol owned 57.43 percent.[32]

04/26/2012—Argentina's Senate approved a bill to expropriate YPF and the Lower House of Congress was expected to vote on the measure making the nationalization official.[33]

2013–2016

No significant expropriations.

No significant privatizations.

NOTES

1. Sergei Guriev, Anton Kolotilin and Konstantin Sonin, "Determinants of Nationalization in the Oil Sector: A Theory and Evidence from Panel Data," *Journal of Law, Economics & Organization*, Vol. 27, No. 2, 2011, p. 301

2. Andrew Inkpen and Michael H. Moffett, *The Global Oil & Gas Industry—Management, Strategy and Finance*, (Tulsa, OK: PennWell Corporation,), 2011, p. 62

3. Ibid.

4. Oil & Gas Journal, "ARCO and Sun Agree to Settle Iranian Claims," (08/24/1992) http://www.ogj.com/articles/print/volume-90/issue-34/in-this-issue/general-interest/arco-and-sun-agree-to-settle-iranian-claims.html.

5. Oil & Gas Journal, "Peru Marks Progress in Privatizing Petroleum," (12/13/1993), http://www.ogj.com/articles/print/volume-91/issue-50/in-this-issue/general-interest/peru-marks-progress-in-privatizing-petroleum.html.

6. Oil & Gas Journal, "Industry Briefs," (10/04/1993) http://www.ogj.com/articles/print/volume-91/issue-40/in-this-issue/general-interest/industry-briefs.html.

7. Andrew Inkpen and Michael H. Moffett, *The Global Oil & Gas Industry—Management, Strategy and Finance*. (Tulsa, OK: PennWell Corporation), 2011, p. 62.
8. Ibid.
9. Oil & Gas Journal, "Industry Briefs," (09/05/1994), http://www.ogj.com/articles/print/volume-92/issue-36/in-this-issue/general-interest/industry-briefs.html.
10. Andrew Inkpen and Michael H. Moffett, *The Global Oil & Gas Industry—Management, Strategy and Finance*. (Tulsa, OK: PennWell Corporation), 2011, p. 62.
11. Ibid.
12. Ibid.
13. BBC News, "Yukos Case against Russia Begins at European Court,"(4 March 2010), http://news.bbc.co.uk/2/hi/europe/8549226.stm.
14. New York Times, "Russian State Oil Company wins Another Yukos Auction," (08/08/2007), http://www.nytimes.com/2007/08/08/business/worldbusiness/08iht-yukos.4.7045853.html?_r=1.
15. Council of Europe, *Parliamentary Assembly, Resolution, 1418* (2005), http://assembly.coe.int/main.asp?Link=/documents/adoptedtext/ta05/eres1418.htm.
16. Oil & Gas Journal, "Marathon cancels Baja California LNG project following site expropriation," (03/02/2004), http://www.ogj.com/articles/2004/03/marathon-cancels-baja-california-lng-project-following-site-expropriation.html.
17. Oil & Gas Journal, "Preliminary Vote Tally Indicates Approval of Bolivian Gas Exports07/20/2004," http://www.ogj.com/articles/2004/07/preliminary-vote-tally-indicates-approval-of-bolivian-gas-exports.html.
18. Oil & Gas Journal, "OPIC Refutes PDVSA Complaints over INTESA Claim," (08/02/2004) http://www.ogj.com/articles/print/volume-102/issue-29/general-interest/opic-refutes-pdvsa-complaints-over-intesa-claim.html.
19. Oil & Gas Journal, "Russian Government to Auction Subsidiary of Yukos on December 19," (11/19/2004), http://www.ogj.com/articles/2004/11/russian-government-to-auction-subsidiary-of-yukos-on-dec-19.html.
20. Oil & Gas Journal, "Watching the World: Bolivia Now, Who's Next?" (05/08/2006), http://www.ogj.com/articles/print/volume-104/issue-18/general-interest/watching-the-world-bolivia-now-whos-next.html.
21. Oil & Gas Journal, "Occidental files claim against Ecuador,"(06/05/2006), http://www.ogj.com/articles/print/volume-104/issue-21/regular-features/ogj-newsletter/ogj-newsletter.html.
22. Oil & Gas Journal, "Bolivia Demands Retail Networks, Refineries," (06/05/2006), http://www.ogj.com/articles/print/volume-104/issue-21/regular-features/ogj-newsletter/ogj-newsletter.html.
23. Oil & Gas Journal, "Revised Alaskan Gas Pipeline Accord Issued," (06/05/2006), http://www.ogj.com/articles/print/volume-104/issue-21/regular-features/ogj-newsletter/ogj-newsletter.html.
24. Alex Nicholson, "Russian Court Declares Yukos Bankrupt," *The Washington Post with Bloomberg*. (August 1, 2016), http://www.washingtonpost.com/wp-dyn/content/article/2006/08/01/AR2006080100500.html.
25. Oil & Gas Journal, "Russian Energy Security," (09/25/2006), http://www.ogj.com/articles/print/volume-104/issue-36/regular-features/editorial/russian-energy-security.html.

26. Oil & Gas Journal, "COMMENT: Sakhalin-2 deal will alter business climate, markets," (01/15/2007), http://www.ogj.com/articles/print/volume-105/issue-3/general-interest/comment-sakhalin-2-deal-will-alter-business-climate-markets.html.

27. Oil & Gas Journal, "Venezuela takes Over Heavy-oil Interests from ConocoPhillips,"(06/27/2007), http://www.ogj.com/articles/2007/06/venezuela-takes-over-heavy-oil-interests-from-conocophillips.html.

28. Oil & Gas Journal, Uruguay attracts interest in its first licensing round," (05/18/2009), http://www.ogj.com/articles/print/volume-107/issue-19/general-interest/uruguay-attracts-interest-in-its-first-licensing-round.html.

29. Oil & Gas Journal, "Crude Takes Largest 1-day Loss since February," (04/19/2010), http://www.ogj.com/articles/2010/04/market-watch-crude6.html.

30. Oil & Gas Journal, "Venezuela Nationalizes H&P Rigs," (07/02/2010), http://www.ogj.com/articles/2010/07/venezuela-nationalizes.html.

31. Oil & Gas Journal, "Madagascar Oil Declares Force Majeure," (03/29/2011), http://www.ogj.com/articles/2011/03/madagascar-oil-declares.html.

32. Oil & Gas Journal, "Repsol Calls Argentina's Nationalization of YPF 'unlawful'," (04/17/2012), http://www.ogj.com/articles/2012/04/repsol-calls-argentinas-nationalization-of-ypf-unlawful.html.

33. Paula Dittrick, Argentina's Senate Approves Bill to Expropriate YPF, Oil & Gas Journal, (04/26/2012), http://www.ogj.com/articles/2012/04/argentinas-senate-approves-bill-to-expropriate-ypf.html.

Appendix III

LIST OF CASES REPORTED BY UNCTAD RELATED TO OIL AND GAS

The 63 cases reported by UNCTAD related to oil and natural gas and related to oil and natural gas and used in the compilation of the statistics used in this book are listed below. For a more complete description of each case and its current legal status see http://investmentpolicyhub.unctad.org/ISDS

LIST OF CASES REPORTED BY UNCTAD RELATED TO OIL AND NATURAL GAS

Ascom S.A and Stati v. Kazahkstan (Case No. 116/2010 SCC)

Tatneft v. Ukraine (PCA OJSC "Tatneft v. Ukraine")

Mercuria Energy v. Poland (SCC, Stockholm Chamber of Commerce)

Mobil Corporation and others v. Bolivarian Republic of Venezuela (ICSID Case No. ARB/07/27)

Mobil Investments Canada Inc. and Murphy Oil Corporation v. Canada (ICSID Case No. ARB(AF)/07/4)

Occidental Petroleum Corporation and Occidental Exploration and Production Company v. Republic of Ecuador and Empresa Estatal Petróleos del Ecuador (ICSID Case No. ARB/06/11)

Renta 4 et al v Russian Federation (SCC Case No 24/2007)

Chevron Corporation and Texaco Petroleum Corporation v Ecuador (PCA Case No. 34877)

RosInvestCo UK Ltd. v. Russian Federation (V 079/2005) (SCC)

Yukos Universal Ltd. v. Russian Federation (PCA Case No. AA 227)

Hulley Enterprises Ltd. v. Russian Federation (PCA Case No. AA 226)

Veteran Petroleum Ltd. v. Russian Federation (PCA Case No. AA 228)

El Paso Energy International Company v. Argentine Republic (ICSID Case No. ARB/03/15)

Occidental Exploration and Production Company v. Ecuador (LCIA Case No. UN3467)

Biederman v. Kazakhstan

Kardassopoulos v. Georgia (ICSID Case No. ARB/05/18)

Ron Fuchs v. Republic of Georgia (ICSID Case No. ARB/07/15)

Azpetrol International Holdings B.V., Azpetrol Group B.V. and Azpetrol Oil Services Group B.V. v. Republic of Azerbaijan (ICSID Case No. ARB/06/15)

National Gas S.A.E. v. Arab Republic of Egypt (ICSID Case No. ARB/11/7)

Mamidoil Jetoil Greek Petroleum Products Societe Anonyme S.A. v. Republic of Albania (ICSID Case No. ARB/11/24)

RSM Production Corporation and others v. Grenada (ICSID Case No. ARB/10/6)

Mohammad Ammar Al-Bahloul v. Republic of Tajikistan, SCC Case No. V (064/2008)

Murphy Exploration and Production Company International v. Republic of Ecuador (ICSID Case No. ARB/08/4)

Caratube International Oil Company LLP v. Republic of Kazakhstan (ICSID Case No. ARB/08/12)

Liman Caspian Oil BV and NCL Dutch Investment BV v. Republic of Kazakhstan (ICSID Case No. ARB/07/14)

Wintershall Aktiengesellschaft v. Argentine Republic (ICSID Case No. ARB/04/14)

Encana v. Ecuador (LCIA Case No. UN3481)

F-W Oil Interests, Inc. v. Republic of Trinidad & Tobago (ICSID Case No. ARB/01/14)

Shell Nigeria Ultra Deep Limited v. Federal Republic of Nigeria (ICSID Case No. ARB/07/18)

Mobil Argentina v. Argentine Republic (ICSID Case No. ARB/99/1)

Total E&P v. Uganda (ICSID Case No. ARB/15/11)

Mobil Investments v. Canada (ICSID Case No. ARB/15/6)

Littop v. Ukraine (SCC)

JKX Oil & Gas v. Ukraine (PCA)

Hanocal & IPIC International v. Korea (ICSID Case No. ARB/15/17)

Cairn Energy Plc v. India

Luxtona v. Russia (Case No. Not Available)

Interpetrol v. Burundi (ICISD Case No. ARB/14/31)

Yukos Capital v. Russia (Case No. Not Available)

MOL Hungarian Oil & Gas Company Plc. V. Republic of Croatia (ICSID Case No. ARB/13/32)

Lone Pine Resources, Inc. v. Canada (ICSID Case No. UNCT/15/2)

Caratube v. Kazakhstan (ICSID Case No. ARB/13/13)

Murphy v. Ecuador (PCA Case No. AA434)

Chevron v. Ecuador (PCA Case No. 2009-23)

Perenco Ecuador Limited v. Republic of Ecuador and Empresa Estatal Petróleos del Ecuador (Petroecuador) (ICSID Case No. ARB/08/6)

Burlington Resources, Inc. and others v. Republic of Ecuador and Empresa Estatal Petróleos del Ecuador (Petroecuador) (ICSID Case No. ARB/08/5)

ConocoPhillips Petrozuata B.V. v. Bolivarian Republic of Venezuela (ICSID Case No. ARB/07/30

Mobil Exploration and Development Inc. Suc. Argentina and Mobil Argentina S.A. v. Argentine Republic (ICSID Case No. ARB/04/16)

AktouPetrol v. Kazakhstan (ICSID Case No. ARB/15/8)

OperaFund v. Spain (ICSID Case No. ARB/15/36)

Nabucco v. Turkey (ICSID Case No. ARB/15/26)

Repsol, S.A. and Repsol Butano, S.A. v. Argentine Republic (ICSID Case No. ARB/12/38)

The Williams Companies, International Holdings B.V., WilPro Energy Services (El Furrial) Limited and WilPro Energy Services (Pigap II) Limited v. Bolivarian Republic of Venezuela (ICSID Case No. ARB/11/10)

Türkiye Petrolleri Anonim Ortakligi v. Republic of Kazakhstan (ICSID Case No. ARB/11/2)

Oil Tanking GMBH v. Bolivia (Permanent Court of Arbitration, PCA)

Pan American Energy LLC v. Plurinational State of Bolivia (ICSID Case No. ARB/10/8)

Universal Compression International Holdings S.L.U. v Venezuela (ICSID Case No. ARB/10/9)

Mærsk Olie, Algeriet A/S v. People's Democratic Republic of Algeria (ICSID Case No. ARB/09/14)

Eni Dación B.V. v. Bolivarian Republic of Venezuela (ICSID Case No. ARB/07/4)

Trans-Global Petroleum, Inc. v. Hashemite Kingdom of Jordan (ICSID Case No. ARB/07/25)

BP America Production Company, Pan American Sur SRL, Pan American Fueguina, SRL and Pan American Continental SRLothers v. Argentine Republic (ICSID Case No. ARB/04/8) (consolidated with ICSID Case No. ARB/03/13)

Pan American Energy LLC and BP Argentina Exploration Company v. Argentine Republic (ICSID Case No. ARB/03/13) (consolidated with ICSID Case No. ARB/04/8)

Pioneer Natural Resources Company, Pioneer Natural Resources (Argentina) S.A. and Pioneer Natural Resources (Tierra del Fuego) S.A. v. Argentine Republic (ICSID Case No. ARB/03/12)

For a complete list of cases brought before an international court or tribunal see http://investmentpolicyhub.unctad.org/ISDS

Appendix IV

LEASE AGREEMENTS AND PRODUCTION-SHARING AGREEMENTS

Figure IV.1 and Figure IV.2 demonstrate the distribution of revenue between an international oil company (IOC) and a host government or its national oil company (NOC) in a Concession or Lease Agreement and a Production-Sharing Agreement. In both examples, the price per barrel is $50. No inference should be made regarding the split in revenue, because the percentages that the parties negotiate will ultimately determine how much of the revenue each will receive.

In a Concession or Lease agreement, the IOC pays all production costs, receives all residual profits and absorbs all residual losses after paying royalties and taxes to the host government or NOC. The advantage to the host government is that it will receive income each quarter in the form of royalties whether the project is operating at a profit or a loss. The host government's royalty receipts are calculated before operating costs and its tax receipts are calculated after operating costs, assuring the state an income stream, even if oil prices decline or operating costs increase. This example is a modern form of concession or lease agreement (royalty/tax system) because it contains a large "draw" on gross revenues in the form of taxes.

Figure IV.2 illustrates a typical production-sharing agreement (PSA) or production-sharing contract (PSC) with a 10 percent royalty, a 60/40 profit split between the host country and the IOC, and a 40 percent tax rate. In principle, royalties should not exist in PSAs, because the host government retains ownership of the oil and gas that is produced, which is reduced only by the IOC's right to the percentage of oil agreed to in the PSA. In this example the IOC's share is $8.86 of each barrel of oil produced on a successful discovery.

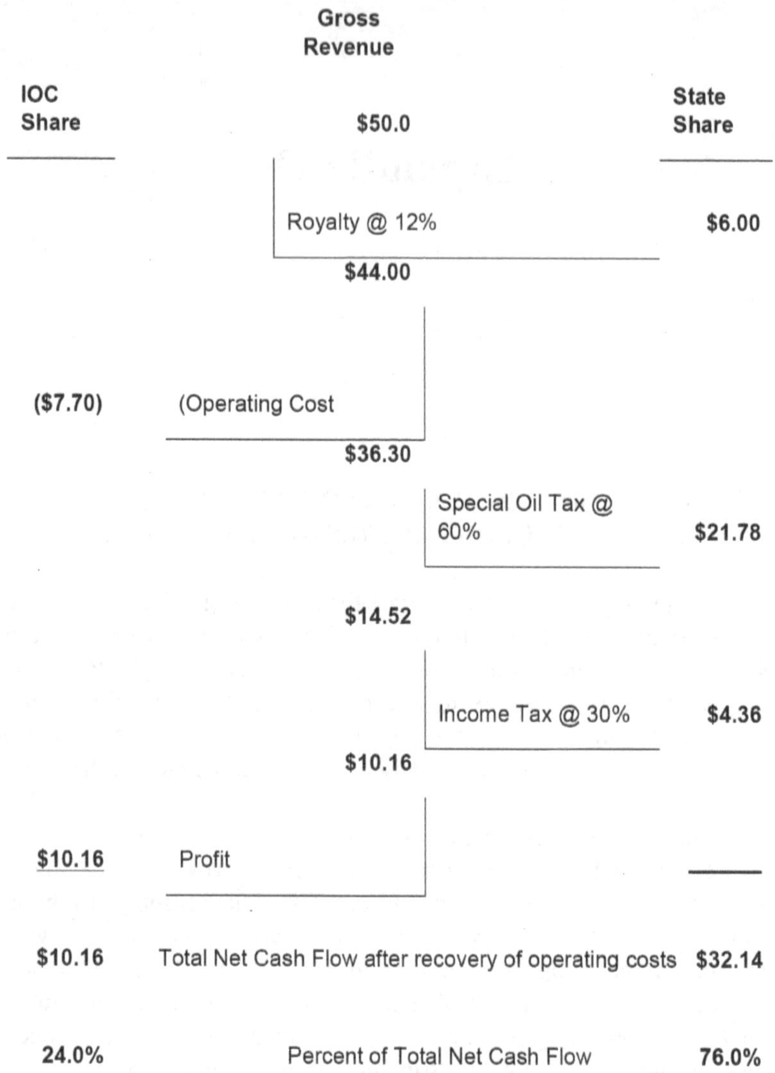

Figure IV.1 Typical Concession or Lease Agreement (Royalty/Tax). *Source*: Adapted from Andrew Inkpen and Michael H. Moffett, The Global Oil and Gas Industry – Management, Strategy and Finance. Chapter 6 (Tulsa, OK: PennWell Corporation, 2011), p. 222

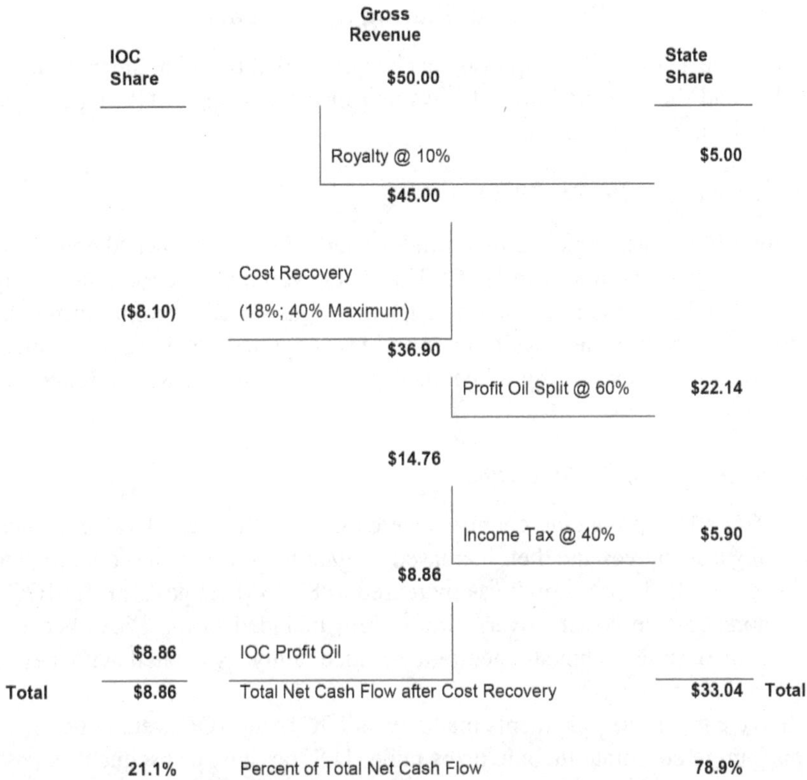

Figure IV.2 Typical Production Sharing Agreement (PSA/PSC). *Source*: Adapted from Andrew Inkpen and Michael H. Moffett, The Global Oil & Gas Industry - Management, Strategy and Finance (Tulsa, OK: PennWell Press, 2011) p. 224.

Nevertheless, royalties are often a component of PSAs.) It is also common for PSAs to provide that as production increases, the proportion allocated to the host government also increases, either in an individual year or cumulatively over the production life of the reservoir. This is usually referred to as a "cumulative production sliding scale."[1, 2] The net financial effect is to raise the effective tax rate on the venture as oil prices increase.

Cost recovery in a PSA is in principle, the deduction of a portion of the oil, called "cost oil," to compensate the IOC for the capital and operating expenses incurred in finding and producing the oil. If, these costs exceed the specified cost recovery limit, however, expenses are not deductible beyond this specified maximum in the current year. The cost recovery limit is typically stated as a percentage of gross revenues earned during the period (40 percent in the example).[3]

The Evolution of Production-Sharing Agreements

PSAs have evolved in response to changing conditions in the petroleum market and the interpretation of laws and incentives for host countries and international oil companies.

First generation (1966 to 1975)

The first PSC was employed by an Indonesian government-owned enterprise and a foreign oil company in 1966.[4] The host government retained ownership of all oil and gas produced, including the oil or gas stored at export terminals. Although there were no royalty rates and taxes applied, the host government was guaranteed revenue by a specified profit split, which was unrelated to cost recovery.[5]

Second generation (1976 to 1983)

By 1976, all oil producing countries were aware of the market value of their oil and gas resources and their increased bargaining power. Consequently, the split of so called "profit oil" was increased to 85/15 (host government/IOC), but more flexible cost recovery limits were included in the PSAs because of the increasing technical (geological) uncertainty associated with newer exploration prospects.[6]

In the early 1970s, payments made by an IOC to an NOC were not considered foreign corporate income taxes under U.S. tax law. Consequently, payments made to the NOC could not be used as tax credits on the IOC's U.S. tax return, when the IOC remitted its foreign profits to the United States. The Saudi government was the first to realize that if the provisions of the PSA were modified, so that the payments were made to the Saudi government rather than to the NOC, the U.S. Internal Revenue Service would classify the payments as foreign taxes, making them eligible for foreign tax credit classification on an IOC's United States tax return. The reclassification of these payments as foreign taxes removed an important obstacle to the major U.S. oil companies signing PSAs in other countries.[7]

Third generation (1984 to 1987)

Minor adjustments to investment tax credits, corporate tax obligations, and corporate tax rates increased the sophistication of PSAs, but did not significantly alter the incentives or disincentives to participate in PSA's.[8]

Fourth generation (1988 to present)

After the substantial decrease in oil prices in the mid-1980s, new PSAs included more flexible terms and conditions to attract IOCs.

The Evolution of Service Contracts

The first service contracts were employed by the Argentine government and its national oil company (YPF) between 1958 and 1961.[9] In a *pure service contract* or risk service contract the international oil company provides some or all the financial capital for exploration and development and is paid by the host government base on a fixed price contract. In a "pure" service contract the international oil company or service company's revenue is based on the activities it performs, like a fixed price construction contract.[10] The Argentine *Frondizi* contracts of the late 1950s, named for Argentina's President at the time, Arturo Frondizi are examples of "pure" service contracts. The international oil companies were required to drill a specific number of wells per year, per exploration block and in exchange received a fixed dollar amount based on a variety of metrics, including meters drilled, wells completed, and ultimately the oil and gas produced per hour.

The terms of a *risk service contrac*t, provide that if the international oil company finds oil or gas, the host country government allows the international oil company to recover its costs by selling its share of the oil and natural gas; however, the international oil company explores for oil and gas at its own risk and expense.[11] If no oil or gas is found, the international oil company bears the cost. The choice of the fiscal regime (concession/lease, production-sharing agreement, pure service contract or risk service contract) determines the allocation of risk and reward between the parties.

NOTES

1. Andrew Inkpen and Michael H. Moffett, *The Global Oil and Gas Industry—Management, Strategy and Finance*. Chapter 6 (Tulsa, OK: PennWell Corporation, 2011), p. 224
2. World Bank Institute, *Guide to Extractive Industries Documents—Oil & Gas*, World Bank Institute Governance for Extractive Industries Program, January 2013, http://wbi.worldbank.org/wbi/Data/wbi/wbicms/files/drupal-acquia/wbi/World%20Bank%20Extractive%20Industries%20Programme%20-%20Oil%20&%20Gas%20Guide.pdf.
3. Andrew Inkpen and Michael H. Moffett, *The Global Oil and Gas Industry—Management, Strategy and Finance*. Chapter 6 (Tulsa, OK: PennWell Corporation, 2011), p. 224.
4. Junseog Yi, "Merits and Demerits of Various Types of Petroleum Contracts," https://webcache.googleusercontent.com/search?q=cache:SErP4i_47isJ:https://www.knoc.co.kr/servlet/Download%3Fnum%3D6%26fno%3D6%26bid%3DDATA1%26callback%3D/sub05/+&cd=1&hl=en&ct=clnk&gl=us.
5. Andrew Inkpen and Michael H. Moffett, *The Global Oil and Gas Industry—Management, Strategy and Finance*. Chapter 6 (Tulsa, OK: PennWell Corporation, 2011), pp. 241–242.

6. Ibid.
7. Ibid.
8. Ibid.
9. Junseog Yi, "Merits and Demerits of Various Types of Petroleum Contracts," https://webcache.googleusercontent.com/search?q=cache:SErP4i_47isJ:https://www.knoc.co.kr/servlet/Download%3Fnum%3D6%26fno%3D6%26bid%3DDATA1%26callback%3D/sub05/+&cd=1&hl=en&ct=clnk&gl=us.
10. Andrew Inkpen and Michael H. Moffett, *The Global Oil and Gas Industry – Management, Strategy and Finance.* Chapter 6 (Tulsa, OK: PennWell Corporation, 2011), p. 226.
11. Ibid.

Appendix V

ACCOUNTING TERMINOLOGY AND CLASSIFICATION SYSTEMS

Under the rules of U.S Generally Accepted Accounting Principles (U.S. GAAP) company expenditures for assets with a useful life of more than one year are classified and recorded as capital additions. This includes property, equipment and infrastructure whether acquired by direct purchase, through the acquisition of another company or participation in a joint venture. The oil and gas industry frequently uses the more inclusive term CAPEX which is the sum of capital additions (which are capitalized on the balance sheet) and exploration expenses (which are expensed on the income statement in the period in which they occur). For example, in the Notes to ExxonMobil's 2015 Summary Annual Report, the company defines CAPEX as:

> "…the combined total of additions at cost to property, plant and equipment and exploration expenses on a before-tax basis from the Summary Statement of Income. ExxonMobil's Capex includes its share of similar costs for equity companies. Capex excludes assets acquired in nonmonetary exchanges (effective 2013) and depreciation on the cost of exploration support equipment and facilities recorded to property, plant and equipment when acquired. While ExxonMobil's management is responsible for all investments and elements of net income, particular focus, is placed on managing the controllable aspects of this group of expenditures. Loans made in exchange for oil or gas are not a part of CAPEX, nor or mergers between companies."[1]

The term "transaction" is another way to group expenditures, used in the news media, but has no meaning in U.S. GAAP or IFRS.

Figure V.1 shows that the financial figures reported by the news media, industry databases and oil and gas companies (IOCs and NOCs) overlap. Global CAPEX

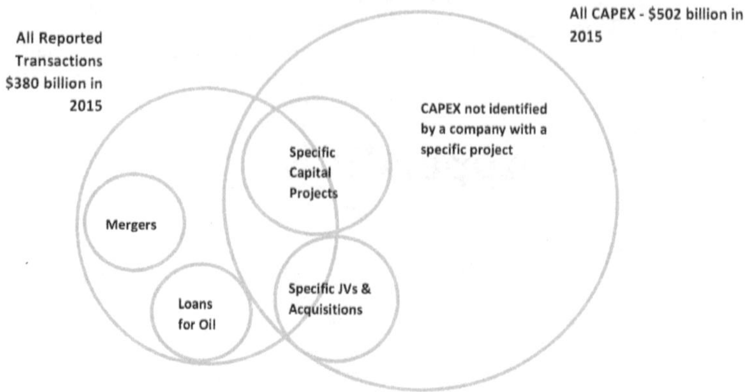

Figure V.1 Relationship between Various Measures of Expenditures 2015. *Source*: Created by the author.

in the oil and gas exploration and development sector were $502 billion in 2015[2] represented by the largest circle. This number includes a large amount of capital expenditures that are not identified publicly with a specific project, as well as those capital expenditures that are identified with a specific project or acquisition. CAPEX does not include loans for oil deals and mergers between companies.

In 2015, the total value of *reported* oil and gas transactions was $380 billion,[3] but this figure excludes all transactions the value of which are not disclosed.[4] Transactions are therefore, a "potpourri" of deals that include specific capital expenditures, acquisitions, joint ventures, loans for oil deals and mergers that are made public in some way, but excludes transactions not identified in the media with a specific expenditure or deal.

These technical distinctions are only distantly related to political and investment risk in the oil and gas industry, but they demonstrate the importance of developing and using a consistent measure of expenditures for analysis. CAPEX is the term most often used in understanding investment trends in the oil and gas industry and the one that is used throughout this book.

NOTES

1. ExxonMobil, "ExxonMobil 2015 Summary Annual Report," http://cdn.exxonmobil.com/~/media/global/files/summary-annual-report/2015_SAR_new.pdf
2. Barclays, "E&P Capex: First Double Dip Since '86," (January 14, 2016), http://www.oilandgas360.com/ep-capex-first-double-dip-since-86/.
3. E&Y, "Global Oil and Gas Transactions Review 2015," http://www.ey.com/gl/en/industries/oil—gas/ey-global-oil-and-gas-transactions-review-2015.
4. E&Y, "Global Oil and Gas Transactions Review 2015," http://www.ey.com/gl/en/industries/oil—gas/ey-global-oil-and-gas-transactions-review-2015.

Appendix VI

LOANS FOR OIL AGREEMENTS

This section presents a list of loans for oil deals from 2006 to the 2016. These transactions usually involve a long-term supply agreement in which, the country producing the oil or gas immediately receives loans from the consuming country for developing natural resources and infrastructure. The country making the loan is repaid by delivery of gas or oil by the country receiving the loan over a specified period.

Loans for oil Deals between 2006 and 2014

2006—(1) China and Nigeria signed a $4 billion agreement for oil and infrastructure projects; an agreement that includes four drilling licenses for China. (2) In a separate transaction, the China National Offshore Oil Corporation (CNOOC) purchased 45 percent of an oil exploration block off the coast of Nigeria for $2.3 billion.[1]

2007—Japan Bank for International Cooperation (JBIC) signed a $3 billion loan deal with the Abu Dhabi National Oil Company (ADNOC) to increase the firm's supply capacity so that the United Arab Emirates (UAE) could continue providing oil exports to Japan.[2]

2008—The Venezuelan Economic and Social Development Bank (BANDES) and PDVSA signed a $4 billion loan for oil deal with the China Development Bank (CDB). The loan was intended to fund infrastructure and other development projects.[3]

2009—(1) Ecuador's national oil company, Petroecuador, signed a $1 billion loan for oil deal with PetroChina, a subsidiary of the China National Petroleum Corp (CNPC) in the form of an advance payment for oil to be delivered later.[4] (2) Venezuela's BANDES and PDVSA signed a $4 billion loan for oil deal with CDB for infrastructure development.[5]

February 18–24. 2009—China and Russia agreed on terms of a loan from the CDB to Russian state oil exporter *Rosneft* for $15 billion and pipeline company *Transneft* for $10 billion, to finance crude oil shipments over a twenty-year period of not less than 241,000 barrels per day (15 million tons per year).[6]

April 18, 2009—State-owned CNPC agreed to lend Kazakhstan's national energy company *KazMunaiGas* $5 billion and join with *KazMunaiGas* to buy Kazakh oil and gas producer *MangistauMunaiGas* from Indonesia's Central Asia Petroleum Ltd. The Export-Import Bank of China agreed to provide an additional $5 billion to Kazakhstan as part of the package.[7]

2010—(1) Ecuador's Petroecuador signed a loan for oil agreement with CDB (20 percent for oil-related investments and 80 percent for infrastructure investments and other discretionary investments).[8] (2) China and Nigeria signed a $23 billion agreement for China to build three oil refineries and a fuel complex in Nigeria.[9]

September 2010—(1) Ghana and China signed project loans and another deal, totaling $15 billion. The China Export Import Bank and the government of Ghana signed a $10.4 billion concessionary loan agreement for various infrastructure projects, payable over twenty years. A separate loan of $3 billion, from the CDB, was intended for Ghana's expanding oil-and-gas sector. The CDB also guaranteed more than $400 million for water projects and what it called e-governance projects in Ghana. (2) In a separate agreement, Ghana signed a deal valued at $1.2 billion with the Chinese company, Bosai Minerals Group, to build a bauxite and aluminum refinery in Ghana over four years. Bosai Minerals will purchase 80 percent of the shares in Ghana Bauxite Co.[10]

2011—(1) Venezuela's PDVSA signed another $4 billion loan for oil deal with CDB for infrastructure construction.[11] (2) Ecuador's government signed a $2 billion loan for oil deal with CDB, 30 percent for oil development and 70 percent for discretionary infrastructure spending.[12]

July 3, 2011—Ecuador was in the final stages of negotiations with a Chinese bank for a $571 million loan; and state oil company, Petroecuador, signed a deal to sell oil to the Chinese energy company, PetroChina.[13]

December 16, 2011—Venezuela's PDVSA signed another $4 billion loan for oil deal with Industrial and Commercial Bank of China (ICBC) for housing development in Venezuela[14] bringing the total amount owed by Venezuela to $30 billion, which was secured by Venezuela's future oil production.[15]

December 3, 2012—PDVSA announced that Citic Group Corp., China's largest state-owned investment company, would acquire a 10 percent stake in the Petropiar heavy-crude project jointly owned with PDVSA and Chevron Corp. It also said that the CDB would spend $4 billion to help boost production in a JV with CNPC. The Chinese bank and the Venezuelan government also agreed to renew a $6 billion bilateral investment fund, of which $2 billion was intended to increase PDVSA's oil production.[16]

April 28, 2012—(1) China agreed to provide South Sudan $8 billion in development loans over the next two years. The loans were to be used for road construction, agriculture, hydroelectricity, infrastructure and telecommunications, which would be built by Chinese companies.[17]

December 24, 2012—OAO Rosneft signed two loan agreements for $16.8 billion with international banks to buy BP Plc's half of TNK-BP. OAO Rosneft also agreed on a prepaid oil supply deal with traders Glencore International Plc and Vitol Group to finance its $55 billion acquisition of TNK-BP.[18]

February 8, 2013—The Japanese Bank for International Cooperation and three Japanese lenders agreed to loan the state-owned oil company of the UAE approximately $3 billion to advance the UAE's upstream development. The Japanese bank made the loan with three other Japanese banks: Bank of Tokyo-Mitsubishi UFJ Ltd., Sumitomo Mitsui Banking Corp. and Mizhuo Corporate Bank Ltd.[19]

May 22, 2013—Essar Energy, the London listed parent of Essar Oil, signed a three-way debt financing deal with CDB, the country's largest overseas lender, and PetroChina, to raise $1 billion of external commercial borrowings (ECBs). The financing cooperation agreement also included a guaranteed product "offtake" by PetroChina.[20]

July 10, 2013—The Nigerian president, Goodluck Jonathan, and the Chinese president, Xi Jinping, signed an agreement to facilitate $1.1 billion in low-interest loans for infrastructure in Nigeria. China was offering Nigeria loans to help fund airport terminals in four cities, roads, a light-rail line for its capitol, a hydro-electric power plant and oil and gas infrastructure. This loan was one part of a $3 billion loan for oil agreement.[21]

October 16, 2013—Turkmenistan's state company *Turkmengaz* and the CDB signed a cooperation agreement for financing the second phase of development at *Galkynysh* field through an undisclosed loan to the Turkmen government. *Turkmengaz* and CNPC also signed a contract for the purchase and sale of 25 billion cubic meters of natural gas to China and a contract for the design and construction of an upstream complex with a capacity to produce 30 billion cubic meters of additional natural gas sales.[22]

October 22. 2013—Russia and China signed twenty-one trade agreements, valued at $85 billion, including a new 100 million-ton oil supply deal with Chinese company SINOPEC. Rosneft agreed to supply China with up to 100 million tons of crude oil over ten years that Rosneft will export through SINOPEC.[23]

February 14, 2014—Venezuela's state oil company PDVSA and Spanish oil firm Repsol completed a $1.2 billion financing agreement to increase output at the *Petroquiriquire* JV. The deal will increase output to approximately 65,000 barrels per day (bpd) from 50,000 bpd.[24]

April 30, 2014—Figures from the China-Latin America Finance Database (a joint effort between the Inter-American Dialogue [a think-tank] and Boston University), indicated that China had committed almost $100 billion to Latin America between 2005 and 2013. The largest amounts have come from the CDB. More than half of China's lending to Latin America has been made to Venezuela. Chinese lenders committed approximately $15 billion in 2013 alone.[25]

May 16, 2014—Venezuela reached a deal with China to improve the conditions of an oil-for-loans deal. Venezuelan economy vice-president Miguel Perez told Reuters that all conditions, including loan time frames, investment amounts and nonfinancial aspects had been approved. [26]

May 16, 2014—Brazilian state-controlled company, Petrobras, signed a contract finalizing a ten-year, $10 billion loan from CDB.[27]

May 21, 2014—Russia agreed to supply China with natural gas for thirty years beginning in 2018 under an agreement that took ten years to negotiate. Russia's state-owned gas company, *Gazprom*, signed the deal with CNPC during the Russian president Vladimir Putin's visit to Shanghai. Under the deal, *Gazprom* will supply 38 billion cubic meters of gas to China each year, with the possibility of increasing shipments to sixty billion cubic meters per year. The Russian Ministry of Energy declined to comment on the price China will pay for the natural gas, but the value of the thirty-year deal was estimated to be $456 billion.[28]

September 2, 2015—President Nicolás Maduro announced a new $5 billion loan from China in his weekly television program.[29]

February 26, 2016—Petroleo Brasileiro SA obtained a $10 billion loan from CDB. The loan is part of a deal to supply crude to China. The quantity of crude that was to be supplied was not disclosed.[30]

NOTES

1. Michail Vafeiadis, "China 'buying out' Africa: Top 5 destinations of Chinese moneyThe Christian Science Monitor," (March 1, 2012), http://www.csmonitor.com/World/2012/0301/China-buying-out-Africa-Top-5-destinations-of-Chinese-money/Nigeria.
2. Jake Simpson, "JBIC Loans $3B to Buoy UAE-Owned Oil Co. Upstream Growth," *Law 360*, New York, (February 8, 2012), http://www.law360.com/articles/413892/jbic-loans-3b-to-buoy-uae-owned-oil-co-upstream-growth.
3. Kevin P. Gallagher, Amos Irwin, Katherine Koleski, "The New Banks in Town: Chinese Finance in Latin America Inter-American Dialogue," (February 2012), http://ase.tufts.edu/gdae/Pubs/rp/GallagherChineseFinanceLatinAmericaBrief.pdf.
4. Ibid.
5. Ibid.

6. John Helmer, "China Loan turns Russian Oil East," *Asia Times on Line,* http://www.atimes.com/atimes/Central_Asia/KB24Ag01.html.

7. Jing Yang and Victoria Ruan, "China, Kazakhstan Sign Loan-for-Oil Deal," *The Wall Street Journal,* (Updated April 18, 2009), http://online.wsj.com/news/articles/SB123996097676128865.

8. Kevin P. Gallagher, Amos Irwin, Katherine Koleski, "The New Banks in Town: Chinese Finance in Latin America Inter-American Dialogue," (February 2012), http://ase.tufts.edu/gdae/Pubs/rp/GallagherChineseFinanceLatinAmericaBrief.pdf.

9. Michail Vafeiadis, "China 'buying out' Africa: Top 5 destinations of Chinese money," *The Christian Science Monitor,* (March 1, 2012), http://www.csmonitor.com/World/2012/0301/China-buying-out-Africa-Top-5-destinations-of-Chinese-money/Nigeria.

10. Reuters, "China Extends Africa Push with Loans, Deal in Ghana," (September 2010)http://ghanaoilonline.org/2010/09/china-extends-africa-push-with-loans-deal-in-ghana/.

11. Kevin P. Gallagher, Amos Irwin, Katherine Koleski, "The New Banks in Town: Chinese Finance in Latin America Inter-American Dialogue," (February 2012), http://ase.tufts.edu/gdae/Pubs/rp/GallagherChineseFinanceLatinAmericaBrief.pdf.

12. Ibid.

13. Reuters, "Ecuador negotiates China bank loan, signs oil deal,"*Taipei Times* (July 3, 2011) http://www.taipeitimes.com/News/biz/archives/2011/07/03/2003507255.

14. Ibid.

15. Los Angeles Times, "China's Venezuela presence Grows with Loan-for-oil Deal," (December 16, 2011), http://latimesblogs.latimes.com/world_now/2011/12/china-venezuela-loan-oil.html.

16. Kelly Hearn, "Venezuelan Oil a Risky Investment for China," *The Washington Times*, (March 12, 2012), http://www.washingtontimes.com/news/2012/mar/12/venezuelan-oil-a-risky-investment-for-china/#ixzz31tLhnzrI.

17. Jared Ferrie, "China to Loan South Sudan $8 Billion for Infrastructure Projects," *Bloomberg News,* (April 28, 2012), http://www.bloomberg.com/news/2012-04-28/china-to-loan-south-sudan-8-billion-for-infrastructure-projects.html.

18. Jake Rudnitsky, "Rosneft Clinches Oil Deal, $16.8 Billion Loans for TNK-BP," *Bloomberg*, (December 24, 2012), http://www.bloomberg.com/news/2012-12-24/rosneft-agrees-to-five-year-oil-supply-deal-with-glencore-vitol.html.

19. Jake Simpson, "JBIC Loans $3B to Buoy UAE-Owned Oil Co. Upstream Growth," *Law 360*, New York (February 8, 2012), http://www.law360.com/articles/413892/jbic-loans-3b-to-buoy-uae-owned-oil-co-upstream-growth.

20. The Economic Times, "Essar Signs Oil-for-loans Deal with CDB, PetroChina to Raise $1 billion ECBs," (May 22, 2013), http://articles.economictimes.indiatimes.com/2013-05-22/news/39445340_1_china-development-bank-vadinar-essar-oil.

21. Associated Press, "Nigerian President in China to Finalize Loan Deals," (July 2013), http://news.yahoo.com/nigerian-president-china-finalize-loan-deals-111649396.html.

22. Tavus Rejepova, "Turkmenistan, China Reach New Energy Deals Field Reports," *CACI Analyst*, October 16, 2013, http://www.cacianalyst.org/publications/field-reports/item/12834-turkmenistan-china-reach-new-energy-deals.html.

23. RT, "Russia and China strengthen Trade Ties with $85 Billion Oil Deal," October 22, 2013, http://rt.com/business/rosneft-china-sinopec-oil-537/.

24. Reuters, "Venezuela Signs $1.2B Loan from Repsol for Joint Oil Deal," http://www.rigzone.com/news/oil_gas/a/131641/Venezuela_Signs_12B_Loan_From_Repsol_For_Joint_Oil_Deal.

25. Law in Action, "News on Environment, Rule of Law, Human RightsThe Thirst for Commodities: Loans-for-Oil Deals," April 30, 2014, http://www.alphabetics.info/international/2014/04/30/loans-for-oil-china-to-latin-america/.

26. Alexandra Ulmer and Corina Pons, "Venezuela Says Better Oil Loans Deal Reached with China," *Reuters,* May 16, 2014, http://www.reuters.com/article/us-venezuela-economy-idUSKCN0Y71VB.

27. Latin American Herald Tribune, "Petrobras Signs $10 Billion Loan Deal with China," May 16, 2014, http://www.laht.com/article.asp?ArticleId=346784&CategoryId=13280.

28. Alanna Petroff, "Russia to Supply China with Gas for 30 year," *CNNMoney,* May 21, 2014, http://money.cnn.com/2014/05/21/news/russia-china-gas/index.html?iid=HP_Highlight.

29. The Economist, "Why China is Lending $5 Billion to Struggling Venezuela," September 2, 2015, http://www.economist.com/news/business-and-finance/21663169-cash-be-invested-oil-will-deliver-dubious-benefits-both-parties-why-china-loaning-5.

30. Sabrina Valle and Denyse Godoy, "Petrobras Gets $10 Billion Chinese Loan in Oil Supply Deal," *Bloomberg*, February 26, 2016, http://www.bloomberg.com/news/articles/2016-02-26/petrobras-gets-10-billion-from-china-to-shore-up-finances.

Bibliography

A Barrel Full, Frade Oil Field, Summary Information, http://abarrelfull.wikidot.com/frade-oil-field. Last accessed on August 20, 2017.

Abs, Hermann and Lord Hartley Shawcross, The Proposed Convention to Protect Private Foreign Investment, *Journal of Public Law*, 9 (1960): 115–124.

Agreement between the Royal Government of Cambodia, Cambodian National Petroleum Agency, Chevron Overseas Petroleum (Cambodia) Limited, Moeco Cambodia Co. Ltd and Woodside South East Asia Pty, Ltd., Article 1.2 Definitions (August 15, 2002).

Agren, David. "Mexico Opens-Up Petroleum Business to Private and Foreign Companies," Special to *USA TODAY*, December 20, 2013, http://www.usatoday.com/story/news/world/2013/12/20/mexico-petroleum-oil-foreign/4152521/.

Alexander's Gas & Oil Connections. "Baku-Tbilisi-Ceyhan Pipeline Company Founded" (August 30, 2002), http://www.gasandoil.com/news/central_asia/06e64ce97241a092b52d5a35d6be6a60.

———. "Angola battles with IMF Accusation of Corruption," *News and Trends Africa*, 7 (22) (November 13, 2002).

Alper, Alexandra and David Alire Garcia. "Mexico's Pemex signs cooperation agreement with Exxon Mobil," *Reuters* (October 2, 2014), http://www.reuters.com/article/2014/10/02/mexico-pemex-idUSL2N0RX16I20141002?feedType=RSS.

AP Worldstream. "Three more oil companies in Venezuela agree to state-controlled joint ventures" (6 October 2005).

Arbogast, Stephen and Praveen Kumar. Financing Large Energy Projects, Chapter 13, in Betty J. Simkins and Russell E. Simkins, *Energy Finance and Economics – Analysis and Valuation, Risk Management, and the Future of Energy*. Hoboken, NJ: John Wiley & Sons, Inc., 2013.

Armbrust, Charles Wiley. "Political Trends in Latin American Oil: Mexico and Venezuela," *Akin Gump, Strauss, Hauer and Feld LLP* (December 2, 2013), http://www.akingump.com/en/experience/industries/energy/speaking-energy/political-trends-in-latin-american-oil-mexico-and-venezuela-1.html.

Arvedlund, Erin E., "A New Twist in Russia's Yukos Oil Affair," *The New York Times* (April 16, 2004), http://www.nytimes.com/2004/04/16/business/a-new-twist-in-russia-s-yukos-oil-affair.html?n=Top%2fReference%2fTimes%20Topics%2fPeople%2fL%2fLebedev%2c%20Platon.

Associated Press. "Coast Guard New Oil Leak from Area Where Rig Exploded, Sank in Gulf; Spill Heads to Coast" (online video) (April 28, 2010), http://www.youtube.com/watch?v=WhUjqi7IlzY.

———. "Venezuela Seizes Oil Rigs Owned by US Company" (June 24, 2010), http://www.cnbc.com/id/.

———. "Nigerian president in China to finalize loan deals" (July 2013), http://news.yahoo.com/nigerian-president-china-finalize-loan-deals-111649396.html.

———. China's CNOOC Drops Bid for Unocal (August 8, 2015), http://www.nbcnews.com/id/8795682/ns/business-oil_and_energy/t/chinas-cnooc-drops-bid-unocal/

Baran, Zeyno. "The Baku-Tbilisi-Ceyhan Pipeline: Implications for Turkey," 103–118, http://www.silkroadstudies.org/BTC_6.pdf.

Barclays. Global 2013, Capital Spending Outlook, Courtesy of Barclays Research Department.

Barstow, David, David Rohde and Stephanie Saul. "Deepwater Horizon Rig Disaster," *The Wall Street Journal* (December 25, 2010) http://www.nytimes.com/2010/12/26/us/26spill.html?pagewanted=all.

———. "E&P Capex: First Double Dip Since," 86 (January 14, 2016), http://www.oilandgas360.com/ep-capex-first-double-dip-since-86/.

Bases, Daniel, Richard Lough and Sarah Marsh. "Argentina, lead creditors settle 14-year debt battle for $4.65 billion," *Reuters*, http://www.reuters.com/article/us-argentina-debt-idUSKCN0W2249 (March 1, 2016).

BBC News, "Venezuela Raises Oil Drilling Tax" (October 11, 2004), http://news.bbc.co.uk/2/hi/americas/3732224.stm.

———. "Venezuela gives Exxon Ultimatum" (October 6, 2005), http://news.bbc.co.uk/2/hi/americas/4544390.stm.

———. "Analysis: Latin America's new left axis" (April 18, 2006), http://news.bbc.co.uk/2/hi/americas/4916270.stm.

———. "Yukos case against Russia begins at European court" (March 4, 2010), http://news.bbc.co.uk/2/hi/europe/8549226.stm.

———. "US Military Joins Gulf of Mexico Oil Spill Effort" (April 29, 2010), http://news.bbc.co.uk/2/hi/americas/8651624.stm.

———. "Argentina to expropriate Repsol oil subsidiary YPF" (April 16, 2012), http://www.bbc.com/news/business-17732910.

Berne Union Statistics 2011–2015 (September 2016), p. 22, http://www.berneunion.org/wp-content/uploads/2016/09/Berne-Union-2016-Charts-and-numbers-for-website.pdf.

Biddle, G., and G. Hilary. "Accounting Quality and Firm-Level Capital Investment," *The Accounting Review,* 81 (2006): 963–982.

Biddle, Gary C., Gilles Hilary, and Rodrigo S. Verdi. "How Does Financial Reporting Quality Relate to Investment Efficiency?" *Journal of Accounting and Economics,* 48 (2009): 2–3, 112–131.

Bierman, Steven and Dina Krennikova, Bloomberg. "Russian Exports set for Record as Europe Competition Grows" (July 3, 2016), http://www.bloomberg.com/news/articles/2016–07–03/russian-oil-exports-set-for-record-as-europe-competition-grows.

Bloomberg News, "BP, Transocean Lawsuits Surge as Oil Spill Spreads in Gulf" (May 1, 2010), http://www.bloomberg.com/apps/news?pid=newsarchive&sid=afgkFlcmKJwU.

Bloomberg New Energy Finance, Fossil Fuel Divestment: A $5 trillion challenge (August 26, 2014), http://about.bnef.com/white-papers/fossil-fuel-divestment-5-trillion-challenge/.

Bloomberg, "Williams Files Arbitration Claim After Venezuelan Seizures" (April 1, 2011), https://www.whatsnextvenezuela.com/tag/united-states/page/20/.

Boykett, Tim, Marta Peirano, Simone Boria, Heather Kelley, Elisabeth Schimana, Andreas Dekrout, Rachel O'Reilly. "Oil Contracts - How to Read and Understand a Petroleum Contract," Times Up Press Industriezeile 33b 4020 Linz. Austria 2012, http://openoil.net/contracts-booksprint/.

BP, "Baku-Tbilisi-Ceyhan Pipeline," http://www.bp.com/en_az/caspian/operationsprojects/pipelines/BTC.html

———. "BTC Celebrates Full Commissioning" (Press release) (July 13, 2006), http://www.bp.com/en/global/corporate/press/press-releases/btc-celebrates-full-commissioning.html.

———. "BP Statistical Review of World Energy" (June 2013), http://www.bp.com/statisticalreview.

———. "BP Statistical Review of World Energy 2014," http://www.bp.com/statisticalreview.

———. "BP Statistical Review of World Energy" June 2015, https://www.bp.com/content/dam/bp/pdf/energy-economics/statistical-review-2015/bp-statistical-review-of-world-energy-2015-full-report.pdf.

———. "BP Statistical Review of World Energy" (June 2016), http://www.bp.com/content/dam/bp/pdf/energy-economics/statistical-review-2016/bp-statistical-review-of-world-energy-2016-full-report.pdf.

Bremmer, Ian, "Managing Risk in an Unstable World," *Harvard Business Review* (June 2005). https://hbr.org/2005/06/managing-risk-in-an-unstable-world

Bronstein, Hugh and Sarah Marsh, "Argentina returns to global debt markets after 15-years" (March 19, 2016), http://www.reuters.com/article/us-argentina-bonds-bids-idUSKCN0XG2W0.

Brown, A. C., *Oil, God, and Gold: The Story of Aramco and the Saudi Kings*. Boston/New York: Houghton Mifflin, 1999.

Bureau of Ocean Energy Management, "BOEM Adjusts Limit of Liability for Oil Spills from Offshore Facilities," *BOEM*, Press Release, http://www.boem.gov/press12112014/.

Bushman, R. and A. Smith, "Financial Accounting Information and Corporate Governance," *Journal of Accounting and Economics*, 31 (2001): 237–333.

Business News Americas, "Silva holds firm on Hydrocarbons law" (May 27, 2002), http://www.bnamericas.com/news/oilandgas/Silva_holds_firm_on_Hydrocarbons_Law.

———. "Seniat to close offices of 2 foreign oilfield operators" (November 4, 2005), http://www.bnamericas.com/news/oilandgas/Seniat_to_close_offices_of_2_foreign_oilfield_operators.

———. "Petrobras Reaffirms Investments Despite Contract Changes" (November 22, 2000), http://www.bnamericas.com/news/oilandgas/Petrobras_reaffirms_investments_despite_contract_changes.

Boue, Juan Carlos, "ConocoPhillips and ExxonMobil v. Venezuela: Using Investment Arbitration to Rewrite a Contract," *International Institute for Sustainable Development* (September 20, 2013), http://www.iisd.org/itn/2013/09/20/conoco-phillips-and-exxon-mobil-v-venezuela-using-investment-arbitration-to-rewrite-a-contract/.

Calatrava, Almudena, "Problems with new Argentine debt payment law", *The Associated Press* (September 13, 2014), http://www.usatoday.com/story/money/markets/2014/09/13/problems-seen-with-new-argentine-debt-payment-law/15580155/.

Calkins, Laurel Brubaker and Jef Feeley, *Insurance Journal* (March 3, 2014), http://www.insurancejournal.com/news/national/2014/03/03/322103.htm.

Calkins, Laurel and Jef Feeley, "BP Must Live with $9.2 Billion Oil-Spill Deal, Court Says," *Bloomberg News* (March 4, 2014), http://www.bloomberg.com/news/2014–03–04/bp-must-live-with-oil-spill-settlement-terms-court-rules-1-.html.

Carlisle, Tamsin, "Venezuela seizes 60 firms," *The National* (May 9, 2009), http://www.thenational.ae/business/energy/venezuela-seizes-60-firms.

Carpenter, J. William, "The World's Biggest State Owned Oil Companies," *Investopedia* (December 17, 2015), http://www.investopedia.com/articles/investing/121715/worlds-biggest-state-owned-oil-companies.asp.

Catholic Relief Services, *Catholic Relief Services Statement on the Chad-Cameroon Oil Pipeline,* http://web.archive.org/web/20140318155730/http://pwypdev.gn.apc.org/sites/pwypdev.gn.apc.org/files/crs_chad_091003.pdf.

CBC News, "Oil Rig Wreck Leaks into Gulf of Mexico," *Associated Press* (April 25, 2010), http://www.cbc.ca/news/world/oil-rig-wreck-leaks-into-gulf-of-mexico-1.911697.

CBS News, "Obama Extends Moratorium on Offshore Drilling" (May 27, 2010), http://www.cbsnews.com/news/obama-extends-moratorium-on-offshore-drilling/.

CFIUS, "Committee on Foreign Investment in the United States," http://www.treasury.gov/resource-center/international/Pages/Committee-on-Foreign-Investment-in-US.aspx.

Chambers, Matt, "Chevron ups cost estimate on Gorgon LNG project to $52 billion." *The Australian*, December 6, 2012, http://www.theaustralian.com.au/business/mining-energy/chevron-ups-cost-estimate-on-gorgon-lng-project-to-52bn/story-e6frg9df-1226530997643

Chevron, "Gorgon Project Update" (August 1, 2014), http://www.chevron.com/media/gorgon/Gorgon-Progress-Update-08012014.pdf.

Chevron Australia, Gorgon Project, https://www.chevronaustralia.com/our-businesses/gorgon. Last accessed on August 20, 2017.

CNN Money, "Russia has just marked 18 months of a deep and painful recession" (August 11, 2016), http://money.cnn.com/2016/08/11/news/economy/russia-economy-recession-six-quarters/.

CNN, "Oil Slick Spreads from Sunken Rig" (video interview) (April 22, 2010), http://www.cnn.com/2010/US/04/22/oil.rig.explosion/.

———. "DOJ Accuses BP of 'Gross Negligence' in Gulf Oil Spill" (September 5, 2012), http://money.cnn.com/2012/09/05/news/companies/bp-oil-spill/.

Corner House, "The Legal Regime for the Baku-Tbilisi-Ceyhan (BTC) Oil Pipeline Project" (March 11, 2011), http://www.thecornerhouse.org.uk/sites/thecornerhouse.org.uk/files/Implications%20of%20NCP%20Final%20Statement%20BP_0.pdf.

Council of Europe, "Parliamentary Assembly, " *Resolution* 1418 (2005), http://assembly.coe.int/main.asp?Link=/documents/adoptedtext/ta05/eres1418.htm.

Courthouse News Service, "Oil Firms May Have Case Against Venezuela" (September 30, 2013), http://www.courthousenews.com/2013/09/30/61616.htm.

Crittenden, Guy. "Understanding the Initial Deepwater Horizon Fire," *HazMat Management* (May 10, 2010), http://www.hazmatmag.com/news/understanding-the-initial-deepwater-horizon-fire/1000370689/?&er=NA.

Crooks, Ed., "BP Refuses to Fund Gulf Oil Spill Studies," *Reuters* (April 20, 2014), http://www.ft.com/intl/cms/s/0/cdf8ba8c-c8a3–11e3–8976–00144feabdc0.html#axzz31WtwsjmY.

Culp, Christopher L. and Paul Forrester, Structured Financing Techniques in Oil and Gas Project Finance, Chapter 21, in *Energy and Environmental Finance Law and Taxation: New Investment Techniques*, ed. Andrea S. Kramer and Peter C. Fusaro. New York: Oxford University Press, 2010.

Deepwater Horizon Claims Center, Claims Administrator Resumes Processing of All BEL Claims, March 1, 2013. http://www.deepwaterhorizoneconomicsettlement.com/docs/BROWNGREER-421750-v3-Alert-Resume_Processing_All_BEL_Claims.pdf

Defotis, Dimitra. "The New MLP Landscape," Barron's (February. 23, 2013), http://online.barrons.com/news/articles/SB50001424052748704103204578314250446774898.

Dickinson, Dan E. and David Wood, "Alaska Tax Reform: Intent Met with Oil," Vol. 107, no. 20 (Part 1 of 2), *Oil & Gas Journal* (May 25, 2009): 20–24.

———. "Alaska Tax Reform: Gas Raises Questions" (Part 2 of 2), *Oil & Gas Journal* (June 1, 2009): 20–26.

Diss, Kathryn and Sue Lannin, "Chevron reveals second Gorgon gas project cost blowout" (December 12, 2013), http://www.abc.net.au/news/2013–12–12/gorgon-gas-project-cost-blowout/5151982.

Dittrick, Paula, "Argentina's Senate approves bill to expropriate YPF," *Oil & Gas Journal* (April 26, 2012), http://www.ogj.com/articles/2012/04/argentinas-senate-approves-bill-to-expropriate-ypf.html.

———. "OPEC World Oil Outlook Sees $95/barrel Oil in 2040," *Oil & Gas Journal*, (December 28, 2015), http://www.ogj.com/articles/2015/12/opec-world-oil-outlook-sees-95-bbl-oil-in-2040.html.

Master Limited Partnerships, Eagle Financial Publications. DividendInvestor.com

Dlhouhy, Jennifer A., "SEC plans oil payment transparency rule by June 2016," *Fuelfix* (October 5, 2015), http://fuelfix.com/blog/2015/10/05/sec-plans-oil-payment-transparency-rule-by-june-2016/.

DW, "Venezuela agrees to pay ExxonMobil for oil nationalization" (October 10, 2014), http://www.dw.de/venezuela-agrees-to-pay-exxonmobil-for-oil-nationalization/a-17985903.

E&Y, "Capitalizing on Opportunities—Private Equity Investment in Oil and Gas," http://www.ey.com/Publication/vwLUAssets/ey-capitalizing-on-opportunities/$FILE/ey-capitalizing-on-opportunities.pdf.

———. "Funding Challenges in the Oil and Gas Sector, Innovative Financing Solutions for the Oil and Gas Sector" (2014), http://www.ey.com/Publication/vwLUAssets/EY-Funding-challenges-in-the-oil-and-gas-sector/$File/EY-Funding-challenges-in-the-oil-and-gas-sector.pdf.

———. "Global Oil and Gas Transactions Review 2015," http://www.ey.com/gl/en/industries/oil---gas/ey-global-oil-and-gas-transactions-review-2015.

———. "A Shrunken Giant" (February 8, 2014), http://www.economist.com/news/business/21595935-british-oil-company-safer-smaller-sadder-and-wiser-its-disaster-gulf.

———. "Why China is lending $5 billion to struggling Venezuela" (September 2, 2015), http://www.economist.com/news/business-and-finance/21663169-cash-be-invested-oil-will-deliver-dubious-benefits-both-parties-why-china-loaning-5.

———. "Venezuela: A Nation in a State" (February 18, 2016), http://www.economist.com/blogs/graphicdetail/2016/02/graphics-political-and-economic-guide-venezuela.

Elder, Miriam. "Russia look to control world's gas prices," *Telegraph* (December 27, 2008). http://www.telegraph.co.uk/earth/energy/gas/3982543/Russia-look-to-control-worlds-gas-prices.html.

Environmental Law Institute, "Deepwater Horizon Oil Spill Litigation Database," http://www.eli.org/deepwater-horizon. (August 10, 2010).

Estevez, Dolia. "Mexico Reverses History and Allows Private Capital into Lucrative Oil Industry" (December 11, 2013), http://www.forbes.com/sites/doliaestevez/2013/12/11/mexico-reverses-history-and-allows-private-capital-into-lucrative-oil-industry/.

Esty, Benjamin C., "Petrozuata: A Case Study of the Effective Use of Project Finance," *Bank of America Journal of Applied Corporate Finance*, 12, No. 3 (1999): 26–42, https://faculty.fuqua.duke.edu/~charvey/Teaching/BA456_2002/Esty_JACF.pdf.

———.*Modern Project Finance – A Casebook*, Chapter 7, John Wiley & Sons: New York (2004).

Exxon Valdez Oil Spill Trustee Council, "Questions and Answers," http://www.evostc.state.ak.us/index.cfm?FA=facts.QA.

ExxonMobil, PNG LNG, (2008). https://pnglng.com/About/Project-overview.

———. PNG LNG, https://pnglng.com/About/Co-venturers.

ExxonMobil, 2015 Summary Annual Report, http://cdn.exxonmobil.com/~/media/global/files/summary-annual-report/2015_SAR_new.pdf

Ezugwu, Anayo, "Petroleum Industry Bill to be Passed Speedily," *Realnews*, (March 11, 2016) http://realnewsmagazine.net/politics/petroleum-industry-bill-to-be-passed-speedily-dogara/.

Feeley, Jef and Allen Johnson, Jr. "BP, Transocean Accused of 'Reckless' Actions in Spill," *Bloomberg* (February 26, 2013), http://www.bloomberg.com/news/2013–02–25/bp-oil-spill-errors-trial-begins-in-new-orleans-court.html.

Ferrie, Jared, "China to Loan South Sudan $8 Billion for Infrastructure Projects," *Bloomberg News* (April 28, 2012), http://www.bloomberg.com/news/2012–04–28/china-to-loan-south-sudan-8-billion-for-infrastructure-projects.html.

Financial Post – Canada, "Venezuela Announces Plan to increase income tax on private oil projects" (April 18, 2005).

Financial Times, "Venezuelan oil law threatens investment" (November 20, 2001).

Forbes, "BP 'Grossly Negligent' In Gulf Spill, Eyes $18 Billion Penalty-And Tax Deduction" (September 4, 2014), http://www.forbes.com/sites/robertwood/2014/09/05/bp-grossly-negligent-in-gulf-spill-eyes-18-billion-penalty-and-tax-deduction/.

Freehills, Herbert Smith, LLP, "ExxonMobil files ICSID arbitration proceedings against Venezuela; energy investors in South America on alert as more World Bank departures threatened" (October 31, 2007), http://www.lexology.com/library/detail.aspx?g=af22d8b6–00d4–4509-aa68–579abb51c7fc.

Gallagher, Kevin P., Amos Irwin, Katherine Koleski, "The New Banks in Town: Chinese Finance in Latin America, Inter-American Dialogue" (February 2012), http://ase.tufts.edu/gdae/Pubs/rp/GallagherChineseFinanceLatinAmericaBrief.pdf.

Gately, Dermot, "How Plausible is the Consensus Projection of Oil below $25 and Persian Gulf Oil Capacity and Output Doubling by 2020?" *Energy Journal,* 22 (2001): 1–27.

———. "OPEC's Incentives for Faster Output Growth," *Energy Journal,* 25 (2004): 75–96.

———. "What Oil Export Levels Should We Expect from OPEC?" *Energy Journal,* 28 (2007): 151–173.

Goldenberg, Suzanne and Dominic Rushe, "BP to pay $4.5bn Penalty over Deepwater Horizon Disaster," *The Guardian* (London) (November 15, 2012), http://www.theguardian.com/environment/2012/nov/15/bp-deepwater-horizon-gulf-oil-spill.

Goldthau, Andreas and Jan Martin Witte, editors, *Global Energy Governance—The New Rules of the Game.* Berlin: Global Public Policy Institute, 2010.

Gourevitch, Peter, *Politics in Hard Times.* Ithaca, NY: Cornell University, 1985.

Grayson, L. E., *National Oil Companies.* Chichester, UK: Wiley, 1981.

Griggs, John Wyeth. "BP Gulf of Mexico Spill Energy," *Law Journal,* 32 (2011): 57–79. http://felj.org/sites/default/files/docs/elj321/14_57_bp_gulf_of_mexico.pdf.

Gulf Oil and Gas, "Oil Spill of Frade Field in Brazil: Special Report," (June 15, 2006). http://www.gulfoilandgas.com/webpro1/projects/3dreport.asp?id=104359.

Guriev, Sergei, Anton Kolotilin and Konstantin Sonin, "Determinants of Nationalization in the Oil Sector: A Theory and Evidence from Panel Data," *Journal of Law, Economics & Organization*, 27, No. 2 (2011): 301.

Harks, Enno, "The International Energy Forum and the Mitigation of Oil Market Risks," Chapter 12, in *Global Energy Governance – The New Rules of the Game*, ed. Andreas Goldthau and Jan Martin Witte, Berlin: Brookings Institution Press, 2010.

Hartley, Peter, Kenneth B. Medlock, III, and Stacy L. Eller, "Empirical Evidence on the Operational Efficiency of National Oil Companies," in *The Changing Role of National Oil Companies in International Energy Markets* May 2011, Vol. 40, no. 3: 623–643, http://link.springer.com/article/10.1007/s00181–010–0349–8#page-1.

Healy, P., and K. Palepu, "Information Asymmetry, Corporate Disclosure, and the Capital Markets: A review of the empirical disclosure literature," *Journal of Accounting and Economics,* 31 (2001): 405–440.

Hearn, Kelly, "Venezuelan oil a risky investment for China," *The Washington Times*, Monday (March 12, 2012), http://www.washingtontimes.com/news/2012/mar/12/venezuelan-oil-a-risky-investment-for-china/#ixzz31tLhnzrI.

Helmer, John, "China loan turns Russian Oil East," *Asia Times On Line,* http://www.atimes.com/atimes/Central_Asia/KB24Ag01.html.

Henkin, Louis, *How Nations Behave.* New York: Columbia University Press, 1979.

Hearn, Kelly, "Venezuelan oil a risky investment for China," *The Washington Times*, (March 12, 2012), http://www.washingtontimes.com/news/2012/mar/12/venezuelan-oil-a-risky-investment-for-china/#ixzz31tLhnzrI ICSID, https://icsid.worldbank.org/apps/icsidweb/cases/Pages/casedetail.aspx?CaseNo=ARB/12/38.

Hydrocarbons technology.com. Baku-Tbilisi-Ceyhan (BTC) pipeline, http://www.hydrocarbons-technology.com/projects/bp/. Last accessed August 20, 2017.

IHS Energy, "CNOOC: Government Drivers" (September 2014).

IRMI (February 13, 2015), https://www.irmi.com/articles/expert-commentary/in-re-deepwater-horizon-additional-insured-questions-resolved.

Iliff, Laurence, "Mexico's Energy Reform Gives Birth to a New Private Oil Company," *The Wall Street Journal* (October 1, 2014), http://online.wsj.com/articles/mexicos-energy-reform-gives-birth-to-new-private-oil-company-1412191958 InflationData.com, http://inflationdata.com/Inflation/Inflation_Rate/Historical_Oil_Prices_Table.asp.

Ingram, David, "Deepwater Horizon: U.S. Ramps up Rhetoric on BP over Oil Spill," *Reuters* (September 4, 2012), http://www.theguardian.com/business/2012/sep/05/deepwater-horizon-us-bp-oil-spill.

Inkpen, Andrew and Michael H. Moffett, *The Global Oil and Gas Industry – Management, Strategy and Finance,* Tulsa, OK: PennWell Corporation, 2011.

Insurance Business, "BP: Insurance policy should cover us for Gulf oil spill" (November 25, 2013), http://www.ibamag.com/news/bp-insurance-policy-should-cover-us-for-gulf-oil-spill-16271.aspx.

Internal Revenue Service Code, I.R.C. § 7704 (2006).

International Centre for the Settlement of Investment Disputes, Case Details, ConocoPhillips Petrozuata B.V., ConocoPhillips Hamaca B.V. and ConocoPhillips Gulf of Paria B.V. v. Bolivarian Republic of Venezuela, (ICSID Case No. ARB/07/30) https://icsid.worldbank.org/ICSID/FrontServlet.

International Convention on Civil Liability for Pollution Damage, 1969, http://www.imo.org/conventions/contents.asp?doc_id=660&topic_id=256.

International Energy Agency, World Energy Outlook 2015, p. 3, https://www.iea.org/Textbase/npsum/WEO2015SUM.pdf.

International Energy Forum, http://www.ief.org/about-ief/history.aspx.

International Finance Corporation, *The Baku-Tbilisi-Ceyhan (BTC) Pipeline Project,* Lessons of Experience (September 2006), No. 2, IFC, World Bank Group, http://www.gcgf.org/wps/wcm/connect/d01d2180488556f0bb0cfb6a6515bb18/BTC_LOE_Final.pdf?MOD=AJPERES&CACHEID=d01d2180488556f0bb0cfb6a6515bb18.

Jacobson, Murrey, "By the Numbers: The Oil Spill and BP's Legal Troubles," *PBS News Hour, the Rundown* (July 9, 2013), http://www.pbs.org/newshour/rundown/gulf-oil-spill-by-the-numbers/.

James, Robert A., *Strategic Alliances between National and International Oil Companies*, Working paper #104: Freeman Spogli Institute for International Studies (October, 2011).

Joint Organizations Data Initiative website, http://www.jodidata.org/about-Jodi.aspx.

Jones, Bart, *Hugo: The Hugo Chavez Story from Mud Hut to Perpetual Revolution*. London: The Bodley Head, 2008.

Jones, Derek A. and Jason B. Kurtz, "Deepwater Horizon: Torrent of oil, flood of insurance issues" (July 23, 2010), http://www.milliman.com/insight/pc/Deepwater-Horizon-Torrent-of-oil--flood-of-insurance-issues/.

Junseog Yi, Merits and Demerits of Various Types of Petroleum Contracts, https://webcache.googleusercontent.com/search?q=cache:SErP4i_47isJ:https://www.knoc.co.kr/servlet/Download%3Fnum%3D6%26fno%3D6%26bid%3DDATA1%26callback%3D/sub05/+&cd=1&hl=en&ct=clnk&gl=us.

Kaufmann, Daniel, Aart Kraay and Massimo Mastruzzi, "The World Governance Project," 2015 Update, http://info.worldbank.org/governance/wgi/index.aspx#home.

King, Leo, "BP £24 billion Lawsuits Claim Contractors Failed to use Modelling Software Properly," Computerworld, UK (April 21, 2011), http://www.computerworlduk.com/news/it-business/3275978/bp-24bn-lawsuits-claim-contractors-failed-to-use-modelling-software-properly/.

King, Rawle O., "Liability and Financial Responsibility Issues Related to Offshore Oil Production, Senate Committee on Energy and Natural Resources" (May 25, 2010), http://www.energy.senate.gov/public/index.cfm/files/serve?File_id=cfe7ede3–9f52–03a2–6432–85f94c70d801.

Kite, Melissa and Nicholas Holdsworth, "Russian bullying over oil is 'a wake-up call," Telegraph (December 24, 2006), http://www.telegraph.co.uk/news/uknews/1537759/Russian-bullying-over-oil-is-a-wake-up-call.html.

Klaes, M, *The New Palgrave Dictionary of Economics*, 2nd Edition, 2008, Abstract. "History of Transaction Costs." ed. Steven N. Durlauf and Lawrence E. Blume. http://www.dictionaryofeconomics.com/article?id=pde2008_T000239&edition=current&q=transaction%20costs&topicid=&result_number=1

Kong, Bo, *China's International Petroleum Policy*. Santa Barbara: Praeger Security International, 2010.

Krauss, Clifford. "In BP Trial, the Amount of Oil Lost Is at Issue," *New York Times* (September 29, 2013), http://www.nytimes.com/2013/09/30/business/energy-environment/bp-trial-in-2nd-phase-to-set-amount-of-oil-spilled.html?pagewanted=all.

Krauss, Clifford and John Schwartz, "BP Will Plead Guilty and Pay Over $4 Billion," *The New York Times*, http://www.nytimes.com/2012/11/16/business/global/16iht-bp16.html?pagewanted=all&_r=0.

Kunzelman, Michael, "Oil Spewing Once Again in the Gulf," *Myrtle Beach News Online Associated Press* (June 24, 2010), http://www.myrtlebeachonline.com/2010/06/24/1550150/oil-spewing-once-again-in-the.html.

———. "Claims against BP contractors dismissed at trial," *The Associated Press* (20 March 2013), http://bigstory.ap.org/article/plaintiffs-rest-trial-over-2010-gulf-oil-spill Retrieved 13 April 2013.

Lambert, R., C. Leuz, and R. Verrecchia, "Accounting Information, Disclosure, and the Cost of Capital", *Journal of Accounting Research,* 45 (2007): 385–420.

Larino, Jennifer, "Federal Judge Rejects BP's Bid to Lower $13.7 Billion Oil Spill Fine," NOLA.com; Times-Picayune (February 19, 2015), http://www.nola.com/business/index.ssf/2015/02/bp_loses_bid_to_cap_oil_spill.html.

Latin American Herald Tribune, "Petrobras Signs $10 Billion Loan Deal with China" (May 16, 2014), http://www.laht.com/article.asp?ArticleId=346784&CategoryId=13280.

———. "U.S. Driller Helmerich & Payne Sues Venezuela over Expropriated Rigs" (June 25, 2014), http://laht.com/article.asp?CategoryId=10717&ArticleId=428308.

Law in Action, "News on Environment, Rule of Law, Human Rights The Thirst for Commodities: loans-for-oil deals" (April 30, 2014), http://www.alphabetics.info/international/2014/04/30/loans-for-oil-china-to-latin-america/.

Leigh, David, "Angolan Oil Millions Paid into Jersey Accounts," *The Guardian* (November 4, 2002).

Levy, Brian, "World Oil Marketing in Transition," *International Organization,* 36 (Winter 1982).

Linde, C. van der, *The State and the International Oil Market: Competition and the Changing Ownership of Crude Oil Assets.* Boston: Kluwer Academic Publishers. 2000.

Lipson, Charles, *Standing Guard: Protecting Foreign Capital in the Nineteenth and Twentieth Centuries.* Los Angeles, CA: University of California Press, 1985.

LOGIC, "Standard Contracts for the UK Offshore Oil and Gas Industry: Service Onshore & Offshore," 2nd edition. (October, 2003). http://www.logic-oil.com/sites/default/files/documents/Services%20Onshore%20and%20Offshore%20Edition%202.pdf.

LOGIC, *Mutual Indemnity and Hold Harmless Deed* (2012), http://www.logic-oil.com/sites/default/files/documents/IMHH%202012%20Deed%20March%20 2014_0.pdf.

Los Angeles Times, "BP Blown-Out Well Finally Killed at Bottom of Gulf of Mexico" (September 19, 2010), http://latimesblogs.latimes.com/greenspace/2010/09/blown-out-bp-well-finally-killed-at-bottom-of-gulf-of-mexico.html.

Los Angeles Times, "China's Venezuela presence grows with loan-for-oil deal" (December 16, 2011), http://latimesblogs.latimes.com/world_now/2011/12/china-venezuela-loan-oil.html.

Lueck, Dean, *The New Palgrave Dictionary of Economics*, 2nd Edition, 2008, "property law and economics," MDL No. 2179, Oil Spill by the Oil Rig "DeepWater Horizon" (August 10, 2010), http://www.laed.uscourts.gov/OilSpill/OilSpill.htm.

MIGA: "Oil and Gas, Securing Oil and Gas Investments, MIGA Briefs" (April 2013), http://www.miga.org/documents/oil&gasbrief.pdf.

Macalister, Terry. "Oligarchs to sue TNK-BP after failing to agree control of company," *The Guardian* (June 11, 2008), http://www.theguardian.com/business/2008/jun/12/bp.oil1?gusrc=rss&feed=business.

Maritime Union of Australia, "Global campaign to target Chevron over Gorgon delays and overruns" (April 9, 2014), http://www.mua.org.au/news-story/3852/.

Market Wired, "Transocean Ltd. Affiliates File Limitation of Liability Petition" (Press release), Transocean (May 13, 2010), http://www.marketwired.com/press-release/transocean-ltd-affiliates-file-limitation-of-liability-petition-nyse-rig-1259747.htm.

Martin, J. Paul. "Chad Cameroon Oil Pipeline Project, a Study Tool and Case Study." Columbia University.

McPherson, C., "National Oil Companies: Ensuring Benefits and Dealing with Systemic Risks," in *The Handbook of Global Energy Policy*, ed. A. Goldthau. Hoboken, NJ: Wiley-Blackwell, 2013.

Milbank, Tweed, Hadley and McCloy, LLP, "Venezuela Withdraws from the World Bank's International Centre for Investment Disputes, Litigation" (January 30, 2012), http://www.milbank.com/images/content/7/2/7277/Venezuela-Withdraws-From-ICSID-1–30–2012.pdf.

Miller, G. J., "The Political Evolution of Principal-Agent Models," *Annual Review of Political Science,* 8, No. 1 (2005): 203–225.

Mohn, Klaus and Petter Osmundsen, "Asymmetry and Uncertainty in Capital Formation: An Application to Oil Investment," *Applied Economics,* Vol. 43, no. 28. (June 2011): 4387–4401.

Mommer, B, *Global Oil and the Nation State.* Oxford, UK/New York: Oxford Univ. Press 2002.

Moody's Investor Service, "Deepwater Horizon Losses Hit Insurers and Reinsurers," *Credit Global Research* (June 30, 2010), https://www.moodys.com/research/Moodys-Deepwater-Horizon-Losses-Hit-Insurers-and-Reinsurers--PR_200382.

Moran, Theodore, "Managing an Oligopoly of Would-Be Sovereigns: Dynamics of Joint Control and Self-Control in the International Oil Industry," *International Organization*, Vol. 41, no. 4 (Autumn, 1987): 575–607.

Mouawad, Jad, "Wildcatters find their niche in the oil industry," *The New York Times*, (May 19, 2007).

Multilateral Investment Guarantee Agency, *World Investment and Political Risk* (Washington, DC: World Bank Group, 2012), 42.

Munich Re, Press Release (September 12, 2010), http://www.munichre.com/en/media_relations/ press_releases/2010/2010_09_12_press_release.aspx.

Munoz, Jeffrey S., "Financing of Oil and Gas Transactions", *Texas Journal of Oil, Gas and Energy Law*, 4, No. 2 (2009): 223–267.

Myers – Jaffe, Amy and Ronald Soligo, "The International Oil Companies," The James A. Baker III Institute for Public Policy (November 2007), http://bakerinstitute.org/files/707/.

National Association of Nationally Traded Partnerships, *National Limited Partnerships 101: Understanding MLPs* (October 4, 2013), http://www.naptp.org/documentlinks/Investor_Relations/MLP_101.pdf.

National Park Service, "Response to Oil on Gulf Island Beaches Continues" (Press release) (June 4, 2010), http://www.nps.gov/aboutus/oil-spill-response.htm.

Newcombe, Andrew and Lluis Paradell, Law and Practice of Investment Treaties: Standards of Treatment. The Netherlands: Kluwer Law International, 2009.

New Era, Center for Energy Economics, Chad-Cameroon Oil Pipeline: A Case Study from New Era, Center for Energy Economics, University of Texas at Austin.

The New Orleans Sun, "BP Found 'grossly negligent' in Gulf of Mexico Oil Spill" (September 2, 2014), http://www.neworleanssun.com/index.php/sid/225407625.

New York Times, "Russian state oil company wins another Yukos auction" (8 August 2007), http://www.nytimes.com/2007/08/08/business/worldbusiness/08iht-yukos.4.7045853.html?_r=1.

Nicholson, Alex, "Russian Court Declares Yukos Bankrupt," The Washington Post with Bloomberg (August 1, 2016), http://www.washingtonpost.com/wp-dyn/content/article/2006/08/01/AR2006080100500.html.

Nolan, Peter A. and Mark C. Thurber, *On the State's Choice of Oil Company: Risk Management and the Frontier of the Petroleum Industry,* Stanford University PESD Working Paper No. 99 (2010), http://pesd.stanford.edu/publications/on_the_states_choice_of_oil_company_risk_management_and_the_frontier_of_the_petroleum_industry/.

North, Douglass C., "Institutions," *Journal of Economic Perspectives*, 5, No. 1 (Winter 1991): 97–112.

North, Douglass C., *Institutions, Institutional Change and Economic Performance.* Cambridge, UK: Cambridge University Press, 1990.

Oberman, Mira, "BP vows to 'vigorously defend' itself at US oil spill trial," *Agence France-Press* (19 February 2013), http://www.globalpost.com/dispatch/news/afp/130219/bp-vows-vigorously-defend-itself-at-us-oil-spill-trialOil & Gas Journal website, http://www.ogj.com/_search?q=expropriation.

Oil & Gas Journal, "ARCO and Sun Agree to Settle Iranian Claims" (August 24, 1992), http://www.ogj.com/articles/print/volume-90/issue-34/in-this-issue/general-interest/arco-and-sun-agree-to-settle-iranian-claims.html.

———. "Industry Briefs" (October 4, 1993), http://www.ogj.com/articles/print/volume-91/issue-40/in-this-issue/general-interest/industry-briefs.html.

———. "Peru Marks Progress in Privatizing Petroleum" (December 13, 1993), http://www.ogj.com/articles/print/volume-91/issue-50/in-this-issue/general-interest/peru-marks-progress-in-privatizing-petroleum.html.

———. "Industry Briefs" (September 05, 1994), http://www.ogj.com/articles/print/volume-92/issue-36/in-this-issue/general-interest/industry-briefs.html.

———. "Marathon cancels Baja California LNG project following site expropriation" (March 02, 2004), http://www.ogj.com/articles/2004/03/marathon-cancels-baja-california-lng-project-following-site-expropriation.html..

———. "Preliminary vote tally indicates approval of Bolivian gas exports" (July 20, 2004), http://www.ogj.com/articles/2004/07/preliminary-vote-tally-indicates-approval-of-bolivian-gas-exports.html.

———. "OPIC refutes PDVSA complaints over INTESA claim" (August 2, 2004), http://www.ogj.com/articles/print/volume-102/issue-29/general-interest/opic-refutes-pdvsa-complaints-over-intesa-claim.html.

———. "Russian government to auction subsidiary of Yukos on Dec. 19" (November 19, 2004), http://www.ogj.com/articles/2004/11/russian-government-to-auction-subsidiary-of-yukos-on-dec-19.html.

———. "Watching the World: Bolivia now, who's next?" (May 08, 2006), http://www.ogj.com/articles/print/volume-104/issue-18/general-interest/watching-the-world-bolivia-now-whos-next.html.
———. "Occidental files claim against Ecuador" (June 05, 2006), http://www.ogj.com/articles/print/volume-104/issue-21/regular-features/ogj-newsletter/ogj-newsletter.html.
———. "Bolivia Demands Retail Networks, Refineries" (June 05, 2006), http://www.ogj.com/articles/print/volume-104/issue-21/regular-features/ogj-newsletter/ogj-newsletter.html.
———. "Revised Alaskan gas pipeline accord issued" (June 05, 2006), http://www.ogj.com/articles/print/volume-104/issue-21/regular-features/ogj-newsletter/ogj-newsletter.html.
———. "Russian Energy Security" (September 25, 2006), http://www.ogj.com/articles/print/volume-104/issue-36/regular-features/editorial/russian-energy-security.html.
———. "COMMENT: Sakhalin-2 deal will alter business climate, markets" (January 15, 2007), http://www.ogj.com/articles/print/volume-105/issue-3/general-interest/comment-sakhalin-2-deal-will-alter-business-climate-markets.html.
———. "Venezuela takes over heavy-oil interests from ConocoPhillips" (June 27, 2007), http://www.ogj.com/articles/2007/06/venezuela-takes-over-heavy-oil-interests-from-conocophillips.html.
———. "Uruguay attracts interest in its first licensing round" (May 18, 2009), http://www.ogj.com/articles/print/volume-107/issue-19/general-interest/uruguay-attracts-interest-in-its-first-licensing-round.html.
———. "Private Equity Investments Crucial to Energy Industry," O&GJ (December 1, 2009).
———. "Crude takes largest 1-day loss since February" (April 19, 2010), http://www.ogj.com/articles/2010/04/market-watch-crude6.html.
———. "Venezuela nationalizes H&P rigs" (July 02, 2010), http://www.ogj.com/articles/2010/07/venezuela-nationalizes.html.
———. "Madagascar Oil declares force majeure" (March 29, 2011), http://www.ogj.com/articles/2011/03/madagascar-oil-declares.html.
———. "Repsol calls Argentina's nationalization of YPF 'unlawful" (April 17, 2012), http://www.ogj.com/articles/2012/04/repsol-calls-argentinas-nationalization-of-ypf-unlawful.html.
———. http://www.ogj.com/content/ogj/en/downloadables/survey-downloads/ogj-100/2013.pdf.
———. http://www.ogj.com/content/ogj/en/downloadables/survey-downloads/ogj-200/2013.pdf.
———. BP's total liability for Deepwater Horizon surpasses $60 billion (July 15, 2016) http://www.ogj.com/articles/2016/07/bp-s-total-liability-for-deepwater-horizon-surpasses-60-billion.html.
———. "Barclays: Global E&P spending to trend up 5% in 2017" (September 7, 2016), http://www.ogj.com/articles/2016/09/barclays-global-e-p-spending-to-trend-up-5-in-2017.html.

Oil Insurance Limited, https://www.oil.bm/.

Olavarría, Leopoldo , Natalija Vojvodic and Sergio Casinelli, "The Venezuelan State reserves assets and services related to hydrocarbon primary activities," http://www.nortonrosefulbright.com/knowledge/publications/58825/the-venezuelan-state-reserves-assets-and-services-related-to-hydrocarbon-primary-activities, May 2011.

Open Secrets .org, Center for Responsive Politics, Influence and Lobbying, https://www.opensecrets.org/lobby/indusclient.php?id=E01.

Organization for Economic Cooperation and Development (OECD), *Council Resolution of 12 October 1967 on the Draft Convention on the Protection of Foreign Property, International Legal Materials,* OECD, 7 (1967): 117–143.

Organization of Petroleum Exporting Countries (OPEC), Nigeria Facts and Figures, http://www.opec.org/opec_web/en/about_us/167.htm.

Organization of Petroleum Exporting Countries (OPEC), http://www.opec.org/opec_web/en/about_us/169.htm.

Organization of Petroleum Exporting Countries (OPEC), Venezuela Facts and Figures, http://www.opec.org/opec_web/en/about_us/171.htm.

Osah, Oviemuno, PIB On Hold Due to Niger Delta Unrest, BizWatchNigeria, (September 6, 2016) http://bizwatchnigeria.ng/pib-hold-due-ndelta-unrest/.

Osipovich, Alexander, "TNK-BP saga raises questions about BP's handling of political risk," *Energy Risk* (March 19, 2013) http://www.risk.net/energy-risk/feature/2253578/tnkbp-saga-raises-questions-about-bps-handling-of-political-risk.

Osmundsen, Petter, Klaus Mohn, Bard Misund, and Frank Asche, "Is Oil Supply Choked by Financial Market Pressures?" *Energy Policy,* 35 (2007): 467–474.

Overseas Private Investment Corporation, *OPIC Annual Report 2013* http://www.opic.gov/sites/default/files/files/OPIC_AR2013_final.pdf.

Overseas Private Investment Corporation, OPIC *Annual Report 2014* https://www.opic.gov/sites/default/files/files/opic-financials-0302.pdf.

Overseas Private Investment Corporation, OPIC *Annual Report 2015* https://www.opic.gov/sites/default/files/files/2015%20Financials_022416–508final.pdf.

Owen, Nick A., Oliver R. Inderwildi and David A. King, "The Status of Conventional World Oil Reserves – Hype or Cause for Concern?" *Energy Policy*, 38, No. 8 (August 2010): 4743–4749.

Parfitt, Tom. "Kremlin attack dog vows to take on Shell in the battle of Sakhalin," *The Guardian* (October 04, 2006), http://www.theguardian.com/business/2006/oct/04/russia.oilandpetrol.

Pearce, Justin, "IMF: Angola's 'missing millions': The sum is three times the amount paid in aid," BBC, Friday (October 18, 2002).

Peralta, Eyder. "Government Files First Criminal Charges in BP Oil Spill," *NPR* (April 24, 2012) http://www.npr.org/blogs/thetwo-way/2012/04/24/151284311/first-criminals-charges-in-bp-oil-spill-filed-against-a-bp-engineer.

Petroff, Alanna, "Russia to supply China with gas for 30 years," *CNNMoney* (May 21, 2014) http://money.cnn.com/2014/05/21/news/russia-china-gas/index.html?iid=HP_Highlight.

Petroleum Agreement Republic of Ghana; Tullow Ghana Limited; Sabre Oil and Gas Limited; Kosmos Energy. Ghana hc, art. 17.7 (Mar. 10, 2006) (Ghana), http://sec.edgar-online.com/kosmos-energy-ltd/s-1a-securities-regis-tration-statement/2011/03/03/section58.aspx.

Pickering, Gordon and Robert Gibb, "Mexico Energy Reform: The Other Story," *Pipeline & Gas Journal*, 243 (August 2016) No. 8, https://pgjonline.com/2016/08/01/mexico-energy-reform-the-other-story/.

Popovic, Neil and Alex Lathrop, "Recovery tactics outlined for foreign takeover losses," *Oil & Gas Journal* (June 25, 2007) http://www.ogj.com/articles/print/volume-105/issue-24/general-interest/recovery-tactics-outlined-for-foreign-takeover-losses.html.

RT, "Russia and China strengthen trade ties with $85 billion oil deal" (October 22, 2013), http://rt.com/business/rosneft-china-sinopec-oil-537/.

Ramseur, Jonathan L., "Liability and Compensation Issues Raised by the 2010 Gulf Oil Spill" (March 11, 2011), http://assets.opencrs.com/rpts/R41679_20110311.pdf.

Raval, Anjli. "Brent Crude Falls Under $50 per Barrel", Financial Times (January 7, 2015), http://www.ft.com/cms/s/0/d04d5e86–9644–11e4-a40b-00144feabdc0.html#axzz3O9da8H3P.

Razawi, Hossein, *Financing Energy Projects in Developing Countries,* PennWell, 2007, Reproduced in Andrew Inkpen and Michael H. Moffett, *The Global Oil and Gas Industry—Management, Strategy and Finance,* Tulsa, OK: PennWell Corporation, 2011.

Rejepova, Tavus, "Turkmenistan, China Reach New Energy Deals Field Reports," *CACI Analyst* (October 16, 2013), http://www.cacianalyst.org/publications/field-reports/item/12834-turkmenistan-china-reach-new-energy-deals.html.

Resnick-Autl, Jessica, "BP Oil-Leak Estimate Doubled by U.S. Science Panel," *BusinessWeek* (June 11, 2010), http://www.bloomberg.com/apps/news?pid=newsarchive&sid=a0uIOSekfLIo.

Reuters, "China extends Africa push with loans, deal in Ghana" (September 2010), http://ghanaoilonline.org/2010/09/china-extends-africa-push-with-loans-deal-in-ghana/.

———. "Ecuador negotiates China bank loan, signs oil deal," *Taipei Times* (July 3, 2011), http://www.taipeitimes.com/News/biz/archives/2011/07/03/2003507255.

———. "Venezuela Signs $1.2B Loan from Repsol for Joint Oil Deal" (February 14, 2014), http://www.rigzone.com/news/oil_gas/a/131641/Venezuela_Signs_12B_Loan_From_Repsol_For_Joint_Oil_Deal.

———. "Struggling to lock in buyers, Chevron's Gorgon casts doubt on global LNG sales model" (August 25, 2014), http://www.reuters.com/article/2014/08/25/chevron-lng-idUSL4N0QQ22X20140825.

———. "US Supreme Court rejects BP challenge to Gulf spill settlement," *The Telegraph* (December 8, 2014), http://www.telegraph.co.uk/finance/newsbysector/energy/oilandgas/11280306/US-Supreme-Court-rejects-BP-challenge-to-Gulf-spill-settlement.html.

Rosen, Raphael, "Why Venture Capital is Weak Fuel for Clean Energy Startups," *Forbes*, http://www.forbes.com/sites/energysource/2013/02/13/why-venture-capital-is-weak-fuel-for-clean-energy-startups/#472eb44050b9.

Ross, Jane. "Florida Coast Suffers First Impact from Oil Spill," *Reuters* (June 4, 2010), http://www.reuters.com/article/2010/06/04/us-oil-spill-florida-idUSTRE6533S820100604.

Rudnitsky, Jake, "Rosneft Clinches Oil Deal, $16.8 Billion Loans for TNK-BP," *Bloomberg*, (December 24, 2012), http://www.bloomberg.com/news/2012–12–24/rosneft-agrees-to-five-year-oil-supply-deal-with-glencore-vitol.html.

Rudolf, John. "Kurt Mix, BP Engineer, Faces First Oil Spill Charges," *Huffington Post* (24 April 2012) http://www.huffingtonpost.com/2012/04/24/kurt-mix-bp-engineer-oil-spill_n_1449316.html.

Rushton, Katherine and Emily Gosden, "Reckless' oil giant could face up to $18bn in Clean Water Act fines for 2010 Disaster" *The Telegraph* (September 4, 2014), http://www.telegraph.co.uk/finance/newsbysector/energy/oilandgas/11075316/BP-faces-up-to-18bn-fines-after-Gulf-of-Mexico-gross-negligence-ruling.htmlBP faces up to $18bn fines after Gulf of Mexico 'gross negligence' ruling.

Rushton, Katherine, "BP Calls for Trial in Deep Water Horizon Case," *The Telegraph* (October 3, 2014), http://www.telegraph.co.uk/finance/newsbysector/energy/oilandgas/11139068/BP-calls-for-retrial-of-Deepwater-Horizon-case.html.

Sanger, David E., Under Pressure, Dubai Company Drops Port Deal, New York Times, (March 10, 2016).

Sayre, Kathryn. "U.S. refuses for now, to stop, BP Oil Spill Payments," *The Times Picayune* (June 9, 2014), http://www.nola.com/business/index.ssf/2014/06/supreme_court_refuses_to_stop.html.

Schmidt, Katherine. "Macondo trial gets under way," *Upstream Online* (NHST Media Group) (February 25, 2013), http://www.upstreamonline.com/live/article1318253.ece.

Schwartz, John. "BP Accuses Texas Lawyer of 'Brazen Fraud' in Workers' Claims Over Gulf Oil Spill," *The New York Times*, (December 17, 2013), http://www.nytimes.com/2013/12/18/us/bp-accuses-texas-lawyer-of-brazen-fraud-in-workers-claims-over-gulf-oil-spill.html?_r=0.

Shaw, Robert W., Jr. "Energy Venture Capital: The New Wave," Arete Corporation, in Energy Venture Capital Best Practices Boston: Aspatore Books, Inside the Minds series, May 2006, 8.

Simons, Marco, "SEC Issues Long-Awaited Transparency Rule for Oil, Gas and Mining," *EarthRights International* (June 28, 2016). https://www.earthrights.org/media/sec-issues-long-awaited-transparency-rule-oil-gas-and-mining.

Simpson, Jake, "JBIC Loans $3B to Buoy UAE-Owned Oil Co. Upstream Growth," *Law 360*, New York (February 8, 2012), http://www.law360.com/articles/413892/jbic-loans-3b-to-buoy-uae-owned-oil-co-upstream-growth.

Snow, Nick, "Halliburton Announces $1.1 Billion Macondo Claims Agreement," *Oil & Gas Journal* (September 3, 2014), http://www.ogj.com/articles/2014/09/halliburton-announces-1–1-billion-macondo-claims-agreement.html.

Soldatkin, Vladimir and Andrew Callus, "Rosneft pays out in historic TNK-BP deal completion," *Reuters* (March 22, 2013), http://uk.reuters.com/article/2013/03/22/uk-rosneft-tnkbp-deal-idUKBRE92K0IX20130322.

Starr, S. Frederick and Svante E. Cornell, "The Baku-Tbilisi-Ceyhan Pipeline: Oil Window to the West," The Central Asia-Caucasus Institute, Silk Road Studies Program (2005) http://www.silkroadstudies.org/BTC.pdf.

Statista, "Statistics Portal." https://www.statista.com/statistics/206764/forecast-for-west-texas-intermediate-crude-oil-prices/.

Steffy, Loren. "New Estimates on Offshore Drilling Ban Costs," *Houston Chronicle*" (June 2, 2010), http://blog.chron.com/lorensteffy/2010/06/new-estimates-on-offshore-drilling-ban-costs/.

———. "BP's Influence Peddling Bears Fruit In Congress," *Huffington Post* (March 12, 2012). http://www.huffingtonpost.com/2012/03/12/bp-oil-spill-gulf-of-mexico-oil-lobbyists_n_1335556.html.

Stein, Sam. "BP's Influence Peddling Bears Fruit in Congress," *Huffington Post* (March 12, 2012), http://www.theguardian.com/business/2012/sep/05/deepwater-horizon-us-bp-oil-spill.

Stevens, P., 2012, *Saudi Aramco: The Jewel in the Crown.* See Victor et al. 2012b, Oil and Governance: State-Owned Enterprises and the World Energy Supply. Cambridge, UK/New York: Cambridge University Press.

SUBSEAIQ "Offshore Field Development," http://www.subseaiq.com/data/Project.aspx?project_id=271.

SUBSEAIQ "Offshore Oil Field Development," http://www.subseaiq.com/data/Project.aspx?project_id=313&AspxAutoDetectCookieSupport=1.

The Australian (December 6, 2012), http://www.theaustralian.com.au/business/mining-energy/chevron-ups-cost-estimate-on-gorgon-lng-project-to-52bn/story-e6frg9df-1226530997643.

The Australian, "Chevron's Gorgon Project 78% Complete" (March 12, 2014), http://www.theaustralian.com.au/business/latest/chevron-eyes-demand-growth/story-e6frg90f-1226852302166?nk=44368b41aa00c57e0356ec94a1c64a47.

The Devil's Excrement, "ExxonMobil Versus PDVSA: Arbitration and Numbers" (January 1, 2012), http://devilsexcrement.com/2012/01/02/exxonmobil-versus-pdvsa-arbitration-and-numbers/.

The Economist, "To the barricades" (November 22, 2001).

The Economic Times, "Essar signs oil-for-loans deal with CDB, PetroChina to raise $1 billion ECBs" (May 22, 2013), http://articles.economictimes.indiatimes.com/2013–05–22/news/39445340_1_china-development-bank-vadinar-essar-oil.

The Telegraph, "TNK-BP: A Troubled History" (July 30, 2014), http://www.telegraph.co.uk/finance/newsbysector/energy/oilandgas/9305157/TNK-BP-a-troubled-history.html.

Thompson, Richard, "Federal judge approves BP Gulf oil spill medical settlement," *The Times-Picayune* (January 11, 2013), http://www.nola.com/business/index.ssf/2013/01/federal_judge_approves_bp_gulf.html.

———. "BP oil spill trial continues as demonstrators note upcoming 3-year anniversary of disaster," *NOLA.com and The Times-Picayune* (April 16, 2013), http://www.nola.com/news/gulf-oil-spill/index.ssf/2013/04/bp_oil_spill_draws_demonstrato.html.

Tienhaara, Kyla, "Foreign Investment Contracts in the Oil & Gas Sector: A Survey of Environmentally Relevant Clauses." *Sustainable Development Law & Policy* Vol. 11, no. 3 (2011): 15–20, 39–40.

The Times-Picayune, "Appeals Court will not re-hear BP Settlement Issue," *Associated Press* (May 19, 2014), http://www.nola.com/environment/index.ssf/2014/05/appeals_court_wont_rehear_bp_s.html.

Tordo, Silvana, Brandon S. Tracy and Noora Arfaa, *National Oil Companies and Value Creation*, *World Bank*, Working Paper 218, Washington, DC, 2011.

Trading Economics, Venezuela Budget Deficit, http://www.tradingeconomics.com/venezuela/government-budget, Last accessed on (October 23, 2016).

Trading Economics, Argentina's Government Budget Deficit, http://www.tradingeconomics.com/argentina/government-budget.

Transocean, "Transocean's Ultra-Deepwater Semisubmersible Rig Deepwater Horizon Drills World's Deepest Oil and Gas Well," Transocean (September 2, 2009), http://globenewswire.com/news-release/2009/09/02/403900/172461/en/Photo-Release-Transocean-s-Ultra-Deepwater-Semisubmersible-Rig-Deepwater-Horizon-Drills-World-s-Deepest-Oil-and-Gas-Well.html.

Trend, "Tankers Carrying Oil Delivered by BTC Pipeline Shipped from Turkey" (April 21, 2014), http://en.trend.az/capital/energy/2265951.html.

Turkish Daily News, "Timeline of the Baku-Tbilisi-Ceyhan Pipeline" (July 13, 2006), http://www.hurriyetdailynews.com/default.aspx?pageid=438&n=timeline-of-the-baku-tbilisi-ceyhan-pipeline-2006–07–13.

UNCTAD, Investment Policy Hub, Investment Dispute Settlement Navigator, http://investmentpolicyhub.unctad.org/ISDS/FilterByEconomicSector.

———. International Investment Agreements, http://investmentpolicyhub.unctad.org/IIA/AdvancedSearchBITResults.

U.S. Coast Guard. "On Scene Coordinator Report on Deepwater Horizon Oil Spill" (September 2011), http://www.uscg.mil/foia/docs/dwh/fosc_dwh_report.pdf.

U.S. Department of Justice, Press Release, U.S. and Five Gulf States Reach Historic Settlement with BP to Resolve Civil Lawsuit over DeepWater Horizon Oil Spill, (October 5, 2016).

Ulmer, Alexandra and Corina Pons, Reuters, "Venezuela says better oil loans deal reached with China," *Reuters* (May 16, 2014), http://www.reuters.com/article/us-venezuela-economy-idUSKCN0Y71VB.

United Nations Conference on Trade and Development, http://investmentpolicyhub.unctad.org/IIA/IiasByCountry#iiaInnerMenu.

———. http://investmentpolicyhub.unctad.org/ISDS/FilterByEconomicSector.

———. Fair and Equitable Treatment – UNCTAD Series on International Investment Agreements II, 2012.

United States Coast Guard, The National Pollution Funds Center, Limits of Liability, https://www.uscg.mil/npfc/Response/RPs/limits_of_liability.asp.

UN News Centre, "Repression in Myanmar, China, Colombia, Venezuela" (March 30, 2004), http://www.scoop.co.nz/stories/WO0403/S00397.htm.

U.S Energy Information Administration, https://www.eia.gov/dnav/pet/pet_move_impcus_a2_nus_epc0_im0_mbblpd_a.htm.

U.S. Energy Information Administration (EIA), "Who are the major players supplying the world oil market?" Energy in Brief, http://www.eia.gov/energy_in_brief/article/world_oil_market.cfm?src=email.

———. *China* (May 14, 2015).

———. Mexico (September 21, 2015), https://www.eia.gov/beta/international/analysis_includes/countries_long/Mexico/mexico.pdf.

U.S. Energy Information Administration, Crude Oil Proved Reserves 2016, https://www.eia.gov/beta/international/data/browser/#/?tl_type=p&f=A&ct=0&c=ruvvvvvfvtvnvv1urvvvvfvvvvvvfvvvou20evvvvvvvvvnvvuvo&ord=SA&pa=0000000000000000000008&tl_id=5-A&cy=2016&v=H&vo=0&so=0&io=0&start=1980&end=2016.

———. *Iraq*, EIA (April 28, 2016).

———. Nigeria, EIA (May 16, 2016).

———. Russia, EIA (July 28, 2015).

———. *Russia*, EIA (November 26, 2013).

———. *Venezuela*, EIA (October 2012).

United States District Court – Eastern District of Louisiana, http://www.stuarthsmith.com/wp-content/uploads/2012/09/7229–2.pdf.

Upstream, "Chevron OK'd for Full Frade Restart" (April 1, 2014), http://www.upstreamonline.com/epaper/article1357081.ece.

USA Today, http://usatoday30.usatoday.com/money/economy/2007–06–26–2141258034_x.htm (June 26, 2007).

USA Today, "Gulf Oil Spill Now Largest Offshore Spill in History as BP Continues Plug Effort" (May 27, 2010), http://www.cacoastkeeper.org/news/gulf-oil-spill-now-largest-in-us-history-as-bp-continues-plug-effort.

USA TODAY (December 20, 2013), http://www.usatoday.com/story/news/world/2013/12/20/mexico-petroleum-oil-foreign/4152521/.

U.S. Department of Justice United States of America v. BP Exploration & Production Inc. et al., Civ. Action No. 2:10-cv-04536, *United States District Court - Eastern District of Louisiana,* http://www.law.uh.edu/faculty/thester/courses/Environmental-Practicum-2014/BP_MOEX_CD_FINAL.pdf.

U.S. Environmental Protection Agency, "Transocean Agrees to Plead Guilty to Environmental Crime and Enter Civil Settlement to Resolve U.S. Clean Water Act Penalty Claims from Deepwater Horizon Incident" (January 3, 2013), http://yosemite.epa.gov/opa/admpress.nsf/2467feca60368729852573590040443d/ea4c5daf4e864d6985257ae80062d9ab!OpenDocument.

Vafeiadis, Michail, "China 'buying out' Africa: Top 5 destinations of Chinese money," *The Christian Science Monitor* (March 1, 2012), http://www.csmonitor.com/World/2012/0301/China-buying-out-Africa-Top-5-destinations-of-Chinese-money/Nigeria.

Valle, Sabrina and Denyse Godoy, "Petrobras Gets $10 Billion Chinese Loan in Oil Supply Deal," *Bloomberg* (February 26, 2016), http://www.bloomberg.com/news/articles/2016–02–26/petrobras-gets-10-billion-from-china-to-shore-up-finances.

van der Linde, C., *The State and the International Oil Market: Competition and the Changing Ownership of Crude Oil Assets.* Boston: Kluwer Academic Publishers, 2000.

Vernon, Raymond, *Sovereignty at Bay: The Multinational Spread of US Enterprises*. New York: Basic Books, 1971.

Victor, David G., "National Oil Companies and the Future of the Oil Industry," *The Annual Review of Resource Economics*, 5 (June 2013): 445–462.

Victor, David G., David R. Hults and Mark C. Thurber, *Oil and Governance: State-Owned Enterprises and the World Energy Supply*. New York, NY: Cambridge University Press, 2012.

Victor, N., and I. Sayfer, *Gazprom: the struggle for power*, 2012, See Victor et al. *Oil and Governance: State-Owned Enterprises and the World Energy Supply*. Cambridge, UK/New York: Cambridge Univ. Press, 2012b, 655–700.

Vitalis, R., *America's Kingdom: Mythmaking on the Saudi Oil Frontier*. Stanford: Stanford Univ. Press, 2006.

Waldie, Paul, "BP turned to Wikipedia to estimate size of spill, U.S. alleges," *The Globe and Mail* (Toronto) (November 15, 2012) http://archive.today/ESugm.

The Wall Street Journal, "Big Oil's Latest Roadblock: Governments Reduce Access to Aid National Champions amid Scramble for Supply" (September 24, 2007), C10.

The Wall Street Journal, "Deepwater Horizon Rig Disaster" (May 19, 2010), http://online.wsj.com/news/articles/SB1000142405274870430230457521388355525958.

Warshaw, C., *The Political Economy of Expropriation and Privatization in the Oil Sector*. See Victor et. al *Oil and Governance: State-Owned Enterprises and the World Energy Supply*. Cambridge, UK/New York: Cambridge Univ. Press, 2012b.

Weaver, Courtney. "Time Line: The Rise and Fall of Yukos," *Financial Times* (July 28, 2914) http://www.ft.com/intl/cms/s/0/f371c836–1645–11e4–93ec-00144feabdc0.html#axzz3O9da8H3P.

Webb, Tim, "BP Boss Admits Job on the Line over Gulf Oil Spill," *The Guardian* (May 13, 2010) http://www.theguardian.com/business/2010/may/13/bp-boss-admits-mistakes-gulf-oil-spill.

Welch, William and Chris Joyner, "Memorial Services Honors 11 Dead Oil Rig Workers," *USA Today* (May 25, 2010). http://usatoday30.usatoday.com/news/nation/2010–05–25-oil-spill-victims-memorial_N.htm.

Weil, "2012–2013 Litigation Wins Report," http://www.weil.com/files/Publication/de949492–1eb8–476b-86f5–295b2dd17d6a/Presentation/PublicationAttachment/e02fcfe9–476f-4cb9–969c-3bb13d2379ae/2012_13_Litigation_Wins_Report.pdf.

Weintraub, E. Roy, The Concise Encyclopedia of Economics, "*Neoclassical Economics*." http://www.econlib.org/library/Enc1/NeoclassicalEconomics.html.

Weisman, Jonathan and Guy Chazan "BP Halts Dividend, Agrees to $20 Billion Fund for Victims," *The Wall Street Journal* (June 17, 2010) http://online.wsj.com/news/articles/SB10001424052748704198004575310571698602094.

What's Next Venezuela, "Timeline of Expropriations," https://www.whatsnextvenezuela.com/media-kit/timeline-of-expropriations/.

Wilkinson, Tracy, "Mexican Political Parties at Impasse over Key Reforms," *Los Angeles Times*. (June 19, 2014), http://www.latimes.com/world/mexico-americas/la-fg-mexican-political-impasse-reform-20140619-story.html.

Williamson, Oliver E., *Markets and Hierarchies, Analysis and Antitrust Implications: A Study in the Economics of Internal Organization.* Free Press, 1983.

Wishon, Jennifer, "BP replaces CEO Hayward, reports $17 billion loss," *CBN News* (July 27, 2010), http://www.cbn.com/cbnnews/us/2010/July/BP-Replaces-Hayward-Reports-17-Billion-Loss-/.

Wood, David A. "Long Term Fiscal Contractual Stability Proves Elusive, Part 1," *Petroleum Review* (February 2005): 38–42. https://www.researchgate.net/publication/274567526_Long-term_fiscal_contractual_stability_proves_elusive_part_1

———. "Long Term Fiscal Contractual Stability Proves Elusive, Part 2," *Petroleum Review* (April 2005): 44–48.

Wood, David, "Petroleum Economics, Risk and Opportunity Analysis," Chapter 10 in *Energy Finance and Economics – Analysis and Valuation, Risk Management, and the Future of Energy,* ed. Betty J. Simkins and Russell E. Simkins. Hoboken, NJ: John Wiley & Sons, Inc., 2013.

Woodruff, David M. "Khodorkovsky's gamble: from business to politics in the YUKOS conflict," *LSE Research Online.* London School of Economics, 2003, http://eprints.lse.ac.uk/3689/1/Khodorkovsky%E2%80%99s_gamble-from_business_to_politics_%28LSERO%29.pdf.

World Bank Institute, *Guide to Extractive Industries Documents – Oil & Gas,* World Bank Institute Governance for Extractive Industries Program (January 2013), http://wbi.worldbank.org/wbi/Data/wbi/wbicms/files/drupal-acquia/wbi/World%20Bank%20Extractive%20Industries%20Programme%20-%20Oil%20&%20Gas%20Guide.pdf.

World Bank, Worldwide Governance Indicators, http://data.worldbank.org/data-catalog/worldwide-governance-indicators.

World's Top Exports, Saudi Arabia's Top Ten Exports (April 1, 2016), http://www.worldstopexports.com/saudi-arabias-top-10-exports/.

Wu, Kang, "China's Overseas Oil and Gas Investment: Motivation, Strategies, and Global Impact," *Oil and Gas Energy Law,* 6, No. 1. (2008): 1–9.

Wustenhagen, Rolf and Tarja Teppo, *What Makes a Good Industry for Venture Capitalists? Risk, Return, and Time as Factors Determining the Emergence of the European Energy VC Market.* University of St. Gallen, IWO Discussion Paper No. 114, October 2004.

Xu, Conglin and Laura Bell, "Worldwide Reserves, Oil Production Post Modest Rise," *Oil & Gas Journal* (December 02, 2013), http://www.ogj.com/articles/print/volume-111/issue-12/special-report-worldwide-report/worldwide-reserves-oil-production-post-modest-rise.html.

Xu, Conglin, Laura Bell, *"OGJ150 suffered huge losses in 2015; capital spending, reserves down,"* "OGJ100 posts lower 2015 earnings, higher oil production Oil & Gas Journal," September 5, 2016, file:///C:/Users/steve/Desktop/ogjournal20160905-dl.pdf.

Y Charts, https://ycharts.com/indicators/average_crude_oil_spot_price.

Yang, Jing and Victoria Ruan, "China, Kazakhstan Sign Loan-for-Oil Deal," *The Wall Street Journal* (Updated April 18, 2009) http://online.wsj.com/news/articles/SB123996097676128865.

Yenikeyeff, Shamil, "BP, Russian billionaires, and the Kremlin: a Power Triangle that never was," *Oxford Energy Comment* (November 2011) http://www.oxfordenergy.org/wpcms/wp-content/uploads/2011/11/BP-Russian-billionaires-and-the-Kremlin.pdf.

Yergin, Daniel, *The Prize – The Epic-Quest for Oil, Money and Power.* New York: Simon & Schuster, 1991.

Yorkville Capital Management, http://www.yorkvillecapital.com/asset-class-overview.aspx.

Zimmermann, Jochen and Jorg R. Werner, *Regulating Capitalism? The Evolution of Transnational Accounting Governance.* New York: Palgrave McMillan, 2013.

INTERVIEWS

DOE, Telephone Interview with a Senior Official in the Office on (September 29, 2014).

ICSID, Telephone Interview with Senior Official at ICSID on (September 23, 2014).

MIGA, Telephone Interview with Senior Official at MIGA on (September 22, 2014).

OPIC, Telephone Interview with Senior Official at OPIC on (September 24, 2014).

Index

About the Author

Steven A. Mucci held various management positions in the oil, gas and chemical industries from 1973 to 2011. In 2015, he earned a Ph.D. in political science at the University of Missouri in St. Louis with a specialization in international political economy. Since 2015, he has been a member of the faculty at the College of St. Benedict and St. John's University in Collegeville, Minnesota, where he has taught courses in Advanced Global Business Strategy, International Business Law, Principles of International Management, Global Enterprise and Globalization, Research Design and Business Ethics. He also holds an MBA in Finance, from George Washington University and a B.S. in Civil Engineering from the University of Missouri.

www.ingramcontent.com/pod-product-compliance
Lightning Source LLC
LaVergne TN
LVHW100728140926
847999LV00001B/5
9781498546126